Praise for *Alex & Me*

"*Alex & Me* is a wonderful read about the close and enduring bonds that developed between a very bright bird and a very motivated researcher during a long-term collaboration. It provides a rare personal and bird's-eye view of the 'ins and outs,' 'ups and downs,' and behind-the-scenes goings-on of scientific research. Irene Pepperberg humanizes science and her dear friend Alex shows that being called a birdbrain is indeed a compliment of the highest order."

—*Marc Bekoff, University of Colorado; author of* The Emotional Lives of Animals, Animals Matter, *and* Wild Justice: The Moral Lives of Animals

"This is a brave book. By insisting on treating Alex as her friend—a friend with whom she laughs and must ultimately grieve over—Dr. Pepperberg takes a stand defying those who insist on confining the issue of consciousness to cold reductionism. I was fascinated to read the anecdotes about the intellectual capacity of parrots, but the best part of *Alex & Me* is the story of their friendship."

—*Mark Bittner, author of* The Wild Parrots of Telegraph Hill

"A wonderful, touching love story that combines humor, history, intimacy, philosophy, and ground-breaking science; a thoroughly delightful read about the thirty-year relationship of Irene and Alex."

—*Joanna Burger, author of* The Parrot Who Owns Me

"Sometimes a single individual changes the world, even if it is a parrot. Together with his tutor, Irene Pepperberg, Alex systematically destroyed the notion—the way he destroyed so many other things—that all that birds can do is mimic human language. Alex clearly had a mind of his own, and a heart to match, as explained in this touching account of scientific perseverance and mutual attachment. Our notion of what a bird is has forever been changed."

—*Frans de Waal, author of* Our Inner Ape

"Everybody who loves animals should read this book. Irene Pepperberg has done pioneering work on communication between people and animals. Alex has proved to the world that birds are much smarter than people think."

—*Temple Grandin, author of* Animals in Translation

Alex & Me

HOW A SCIENTIST
AND A PARROT DISCOVERED
A HIDDEN WORLD OF ANIMAL
INTELLIGENCE—AND FORMED A
DEEP BOND IN THE PROCESS

Irene M. Pepperberg

COLLINS
An Imprint of HarperCollins Publishers

HarperCollins books may be purchased for educational, business, or sales promotional use. For information, please write: Special Markets Department, HarperCollins Publishers, 10 East 53rd Street, New York, NY 10022.

FIRST EDITION

Designed by Leah Carlson-Stanisic

Library of Congress Cataloging-in-Publication Data

Pepperberg, Irene M. (Irene Maxine).

Alex & me: "how a scientist and a parrot discovered a hidden world of animal intelligence—and formed a deep bond in the process"/Irene M. Pepperberg.—1st ed.

p. cm.

ISBN-13: 978-0-06-167247-7

1. African gray parrot—Behavior. 2. Cognition in animals. 3. Animal communication. 4. Human-animal relationships. I. Title. II. Title: Alex and me.

QL696.P7P457 2008

636.6'8650929—dc22

2008021960

08 09 10 11 12 ov/rrd 10 9 8 7 6 5 4 3 2 1

Contents

||||||||||||||||||||||||

Chapter 1

||||||||||||||||||||||||

My Wonderful Life *Moment*

How much impact could a one-pound ball of feathers have on the world? It took death for me to find out. And so I write the story of a particular bird's life, but it must begin at the end.

"*Brainy Parrot Dies*, Emotive to the End," ran a *New York Times* science section headline on September 11, 2007, the day after our press release announcing Alex's passing. "He knew his colors and shapes, he learned more than 100 English words," wrote Benedict Carey, "and with his own brand of one-liners he established himself in television shows, scientific reports and news articles as perhaps the world's most famous talking bird." Carey quoted my friend, colleague, and expert on dolphin and elephant communication, Diana Reiss: "The work revolutionized the way we think of bird brains. That used to be a pejorative, but now we look at those brains—at least Alex's—with some awe."

I found myself saying much the same thing in the newspaper, magazine, radio, and television interviews that overwhelmed me those first few days. People would ask, "What is all the fuss about, why was Alex so special?" and I'd say, "Because a bird with a brain the size of a shelled walnut could do the kinds of things that young children do. And that changed our perception of what we mean by 'bird brain.' It changed the way we think about animal thinking." That was the scientific truth I had known for many years, and now the idea was beginning to be accepted. But that didn't help me with the personal devastation.

Friends drove up from Washington, D.C., that first weekend to ensure I would not be alone, that I would eat and at least try to rest. I functioned each minute, hour, day on automatic pilot, doing whatever was necessary, deprived of sleep, torn by grief. And all amidst this very public outpouring. I was aware of it, of course, yet not fully aware, not then, anyway. I was cognizant of the gathering acclaim, inevitably so because of this endless stream of interviews. But it seemed to involve someone else, or at least had an unreality to it. The phone would ring and I'd click into "interview mode," responding as I had many other times when something Alex had done occasioned a media blitz, responding in a professional manner to the inquiries. This time, however, I'd fall apart until the next call.

Pictures of Alex appeared on CNN, in *Time* magazine, and in scores of other places across the country. National Public Radio ran a story on *All Things Considered*: "Alex the Parrot, an Apt Student, Passes Away." *ATC*'s host, Melissa Block, said, "Alex shattered the notion that parrots are only capable of mimicking words." Diane Sawyer did a two-and-a-half-minute segment on ABC's *Good Morning America*—long for morning television, I'm told. "And now I have a kind of obituary," she began, "and I want to inform the next of kin about a death in the family. And, yes, the next of kin would be all of us." She said that Alex had been a kind of bird genius, "opening new vistas on what animals can do." She aired a video that showed Alex answering questions about the color, shape, and number of objects, and so on. The video landed on YouTube. The previous day, CBS anchor Katie Couric devoted more time to Alex's life and death than to major political stories.

Two days later, the prominent British newspaper, *The Guardian*, wrote, "America is in mourning. Alex, the African Grey parrot who was smarter than the average U.S. president, has died at the relatively tender age of 31." The story was spreading around the world, eventually to Australia. Robyn Williams, from the Australian Broadcasting Corporation's radio *Science Show*, interviewed me, the second time we'd talked about Alex and his achievements. The first time, five years earlier, we'd talked about what other feats Alex might achieve in his future. Not this time.

I was told that the *New York Times* article had been the most e-mailed story of the day, even while General David Petraeus was testifying in Washington, D.C., on Iraq. A second *New York Times* article, on September 12, in its Editorial Notebook section, was titled simply "Alex the Parrot," by Verlyn Klinkenborg. This piece was a little more philosophical than most. "Thinking about animals—and especially thinking about whether animals can think—is like looking at the world through a two-way mirror," Klinkenborg began. "There, for example, on the other side of the mirror, is Alex. . . . But looking at Alex, who mastered a surprising vocabulary of words and concepts, the question is always how much of our reflection we see." The article ended: "The value [of the work] lies in our surprise, our renewed awareness of how little we allow ourselves to expect from the animals around us." A lovely piece, another acknowledgment. But it still felt unreal.

Even Jay Leno had a crack at Alex, on his late-night TV show. (A friend told me about it; I don't have a working TV.) "Sad news: a thirty-year-old parrot by the name of Alex, who had been used by researchers at Harvard University to study how parrots communicate, has died," said Leno. "I believe his last words were, 'Yes, I want a cracker!' " He went on, "This parrot was very intelligent. They say he knew over one hundred words. They say his intelligence was somewhere between a dog and Miss Teen South Carolina." Sigh.

By now every major newspaper had covered Alex's death, noting his remarkable cognitive skills and our breakthrough work together. Even the venerable British science journal *Nature* wrote about it in "Farewell to a Famous Parrot." "Pepperberg has published dozens of scientific papers about Alex's verbal, mathematical and cognitive abilities," noted David Chandler, "and the two have appeared on a wide variety of television programmes and popular press stories." Chandler continued, "In the process, they have transformed people's understanding of the mental abilities of non-human animals." (A bittersweet irony here: when I started working with Alex three decades earlier, a paper I submitted to *Nature* was summarily dismissed without review—as was another I had submitted more recently.)

If, in retelling this outpouring of public recognition, I seem oddly absent, it is because in truth I was. Inasmuch as I was aware of article after article—and friends were assiduous in sending them to me—I continued to let them and their message wash over me. Yes, I was busy with the issue of facing and surviving each new day, busy being interviewed, busy with the lab. At the same time, I could hardly hear what was being said. I had for years been hoping that Alex's achievements would be fully acknowledged, and now it had happened, but I couldn't see it clearly, hear it clearly. Not immediately, anyway.

When, a little more than a week after Alex left me, the *New York Times* did a *third* article, "Alex Wanted a Cracker, but Did He *Want* One?" I began to take notice. George Johnson, a senior science writer, beautifully described the research, and addressed the issue of intention, implicit in the article's title. In the United States, the *Times* is a touchstone for public recognition, whether in politics, the arts, or the sciences. And here was Alex, appearing *three* times within a week in the paper of record. *Hmm*, I thought. *Maybe there's something to all this?*

Then, a few days later, a friend called. "Irene, you are never going to believe this. Alex is in the *Economist*!" She was right. I wasn't going to believe it. The *Economist* is probably the world's preeminent weekly magazine on politics, finance, and business. Each week it has a one-page obituary of some notable dignitary. In the September 20 issue, Alex was that dignitary. Alex's death, said the article, brought to an end "a life spent learning complex tasks that, it had been originally thought, only primates could master." The obituary went on to say that "by the end [of the study] Alex had the intelligence of a five-year-old child and had not reached his full potential." Not reached his full potential—how true, how tragically true.

Given that in the weeks prior to Alex's obit the *Economist* had run Luciano Pavarotti's, Ingmar Bergman's, and Lady Bird Johnson's, I knew just how big an honor it was for Alex to be on this obituary page. It really caught my attention.

In the days and weeks following Alex's death I was roiled by multiple tsunamis of surprise, around me and within me, while struggling to deal with the practical matters, answering phone calls, making arrangements, and much more, because of who Alex was. And my mind was desperately churning: *What's to become of the lab? What's to become of the research? What's to become of everything we've created? What's to become of me?*

I felt swept up in the kind of speeded-up, whirling, swirling cloudscape that one sometimes sees in movies. Except that the concept of the cloudscape went beyond the physical image of chaos to a reality that turned upside down everything that I knew, or thought I knew, about my life.

And *surprise* was indeed the correct term, even if too simple a word to impart the true weight of its message. The sense of loss, grief, and desertion that tore viscerally at my heart and soul at the passing of my one-pound colleague and companion of three decades was of a degree and intensity I had never anticipated, nor could have imagined. A huge torrent of love and caring, assiduously kept in check by a solid dam for all that time, suddenly burst through; the liberated flood of emotions swept all reason before it. I have never felt such pain nor shed more tears. And hope never to again.

Now, I said that a great torrent of emotions had been assiduously kept in check for three decades, as if by some third party I'd hired to do the job, some outside contractor, Emotion Controllers, Incorporated. Of course, the one doing the controlling all that time had been me. My decision. My plan. My implementation. But I had become so good at implementing my plan of emotional distance that this profound torrent of feelings that was the subterranean currency between Alex and me lay out of sight, invisible even to me, beyond the rugged mountains of the cause of scientific objectivity. Mostly invisible, anyway. Mostly out of sight.

I realize that what I just said might make little or no sense to many people, might seem a little Tolkienesque, even. But, in truth, there is something a little Tolkienesque about the thirty-year journey that Alex and I undertook together: the struggles, the initial triumphs, the setbacks, the unexpected and often stunning achievements. And, of course, the premature, final separation. All will unfold, including the rationale for erecting the emotion dam, in the following pages of *Alex & Me*. But my point here is that the internal tsunami I experienced after Alex left me and traveled to what many call "the Rainbow Bridge" was the seismic shock of previously unexpressed emotion, emotion now set free. Yes, I had always cared about Alex, always referred to him as my close colleague, and always treated him with the kind of affection and respect one would have for any close colleague. But I also always had had to maintain my distance, to report the science objectively. Now there would be no more science, at least with Alex, and I could no longer maintain that objectivity.

The external tsunami was no less surprising. As I weathered the media onslaught, e-mail condolences began to arrive. A trickle at first, but within a few hours it became a torrent, then a flood. Jaimi Torok, our Web master, set up a separate condolence site, Remembering Alex, so as not to overwhelm the server of the foundation that supported my research with Alex. More than two thousand messages were posted within a week, three thousand by the end of the month; my own e-mail was awash with just as many. Some were from people I knew, such as former students; I was comforted by hearing how their time with Alex and me had been so important to them and had helped steer them in their lives. Some were from people who had visited the lab just once and wanted to remember that special occasion and share it again. But most were from complete strangers, people who were simply moved to write. Many were "parrot people," of course, but not all. And what they wrote truly astonished me, another of those tsunamis of surprise.

Of course, I wasn't totally unaware of Alex's impact. Soon after Alex and I started working together, I began to be invited to talk at parrot clubs and conferences, to tell people what I had discovered in my work with Alex.

Parrot owners are passionate about their birds, and what I told these parrot people about Alex vindicated what they *knew* about their own birds. They could tell their skeptical friends, "See, I told you so!" This theme was prominent in the condolence Web site. Let me give a few examples:

"It goes without saying that Alex and Irene pushed into realms others thought at minimum silly and otherwise absurd, but we Grey folk know better," wrote Laurence Kleiner, a pediatric neurosurgeon at Children's Hospital in Dayton, Ohio. He is also president of Wings Over the Rainbow, a rescue/rehab organization for abandoned or unwanted birds. "Alex was the beacon and Irene the charge to make it happen; to show the world how truly remarkable our avian friends are. Displaying so elegantly the talents and feelings hithertofore attributed only to humans; how egocentric of us as a species . . . Alex will be remembered always by thousands."

"I cried like a baby when I heard about Alex's untimely death," wrote Linda Ruth. "As a biologist, veterinarian and lifelong bird owner, I found Alex's accomplishment to be a remarkable demonstration of the intelligence and abilities many animals have. . . . Using Alex [as an example] I have been able to convince many skeptics that the gulf between humans and animals is not nearly as deep as once thought."

"As a co-owner of an exceptional Grey, I am devastated by this shocking news," wrote a male financial executive in New England. "I am not an overly sensitive or maudlin person, but I had to leave work for a while upon learning of Alex's death, and my eyes have been welling up at various points throughout the day. My deepest sympathies to all of you who have worked so hard with this inimitable, surpassingly beautiful creature."

"Gandhi once said 'Be the change you want to see in the world,'" wrote Karen Webster, the managing director of the Anchorage chapter of Parrot Education & Adoption Center. "Irene and Alex were that change. One woman working with one (in the beginning, anyway) gray ball of personality has helped bring greater understanding and thus vast improvement into the lives of parrots worldwide. Quite a legacy."

As you will come to see in the following chapters, science was what drove me so hard over the years as I tried to understand the workings of brains of creatures other than ourselves, "lowlier" than ourselves. Many people wrote beautifully about this aspect of our lives, and how the science was bound together with Alex's emotional impact:

"I taught an undergraduate course in animal behavior a few years ago and introduced the class to Alex, showing the infamous PBS video with Alan Alda," said Deborah

Duffy, a researcher in animal behavior at the School of Veterinary Medicine, University of Pennsylvania. "They were amazed! Alex made a strong impression on my students and was the most commonly cited answer to essay exam questions about animal cognition. He was an ambassador of non-human animals, showing us that you don't need to have a brain that looks human to possess complex cognitive abilities. His passing is a sad loss to the scientific community, to education, to animal lovers, and to the world. We will miss him."

"I wish to compliment you, Dr. Pepperberg, on the courage it took for your initial proposal and for sticking it out for each raising of the bar," said David Stewart, an economist in Washington, D.C., and product of a family with many, many pets over the years. "As for the remaining skepticism regarding your work, I regard it as narcissism about the uniqueness and specialness of humanity. . . . Over time I believe it will be widely recognized that whatever defines humans is only a matter of a different degree, rather than a binary having it or not having it. Your work has contributed greatly to this. . . . With sympathy, tears, and gratitude."

Susanne Keller, owner of a Grey, in Alaska, wrote: "Now and then I believe we are sent a special messenger, one who has been sent to teach us something when the time is right. . . . Then along came Alex. A small gray bird. I don't believe that either Dr. Pepperberg or Alex could have ever envisioned what a monumental task they had been given. Nor would they have ever known what a profound difference they would make. . . . Alex was truly a gift to us. He and Dr. Pepperberg were meant to be a team. They needed each other to teach those lessons. . . . Alex, you are one of those rare beings that have made such a positive change in this world."

Of course, most people had never met Alex, perhaps didn't have a bird of their own, but nevertheless were moved by him in some way, helped by him in some way. One letter is especially poignant:

"This is a true story," it began. "Back in the late 1980s, a woman in her mid-thirties was diagnosed with a complicated heart arrhythmia that couldn't be fixed, could barely be controlled, and was severe enough that every incident could possibly be fatal. As a result, she could do very little. She lost what seemed like almost everything—her hope to have a baby, her career, the ability to do some of the simplest activities. Because her husband traveled for his job, she was alone quite a lot. As someone whose life had been full of activities and goals, the sudden emptiness of the future was unbearable. She often looked at the medication that kept her alive and thought about just not taking it.

"Then she read an article about an amazing parrot named Alex and his equally amazing mentor, Dr. Irene Pepperberg. To the woman, who loved animals a great deal, the work that Alex and Irene were doing together was so interesting, so unique and important that she had to know as much about it as she could. To think that a parrot could not only speak, but could know—could absolutely understand—what he heard and what he said, was a miracle in itself to the woman who had stopped believing in miracles. So, for the first time since her illness, the woman set herself a goal: to experience for herself the miracle that Alex and Irene were proving to the world of science.

"I know this story is true because it's mine. Nearly two decades later, after an experimental surgery and its disabling complications, I'm still here, still following the work done by The Alex Foundation. My own parrots (including, of course, a sixteen-year-old African Grey) are still a miracle to me with every word they speak. They are still my lifeline.

"But it was Alex and Irene who threw that lifeline out for me to catch, so many years ago.

"To Irene and all members of The Alex Project, my heartfelt prayers go out to you. Be sure that Alex will never be forgotten by those of us whose hearts have been touched by this amazing little soul." The message was signed Karen "Wren" Grahame. I later discovered that this was the same Wren who had been sending a $10 check to The Alex Foundation every month for years. I had never before known her backstory.

"I was never fortunate enough to meet Alex or Dr. Pepperberg but I feel as if I have known them for along time," wrote Denise Raven, of Belton, Missouri. "This breaks my heart; I have a lump in my throat, and a terrible sense of loneliness inside. It is amazing how deeply this little guy touched so many lives. I thank God for being able to have Alex, Dr. Pepperberg, and The Alex Foundation be a part of my life. I lost my only child 4 years ago and I have to say that this pain of losing Alex hits me as hard as losing my child. I just can't seem to shake the pain. All I can say is that . . . you made this world a better place and you are so missed here."

"My heart feels broken today," said Patti Alexakis. "Alex stole my heart many years ago. He was a little prince—a bright shining star. Godspeed, Alex. You are a much beloved Grey and will forever remain in my heart and many others. I've a started an Internet memorial candle for you, Alex—and for your loving humans. Please light a candle if you wish."

Bill Kollar's tribute was, I have to say, one of the more

unusual and loving ones that Alex received. Kollar is an engineer in northern Virginia and is head of a band of church bell ringers. "On September 16, the bells of Calvary Church in Frederick, Maryland, rang to Alex's memory," he wrote in his e-mail. Kollar owns an African Grey and knew of Alex; so too did his band of ringers, apparently. "One of my professional rules is that you should always ask people to do what they are good at," he later said. "As for bell ringers, when a significant event occurs, such as Alex's death, we ring." And so they did, a forty-three-minute-long quarter-peal combination on the church's six newly hung bells, an unearthly sound winging across the countryside—a beautiful thought. I worry that I never managed to send him a thank-you; even now, I can't remember to whom I wrote during those dark days.

"I cannot begin to tell you how deeply I feel for you," wrote Mother Dolores Hart. "I am a member of the Abbey of Regina Laudis community of Benedictine contemplative nuns, in Bethlehem, Connecticut. We have a Grey who we love so much and who has kept me in close awareness of you and your work. Your sudden loss is heartbreaking. We will keep you in our prayers and our love of these wondrous creatures who have shown us something more of God than we could have ever believed possible." Mother

Hart is the abbey's prioress, a former Hollywood actress who had the distinction of playing opposite Elvis Presley in two films and starred in the 1960 classic *Where the Boys Are*. She left the glamour of film and entered the Abbey of Regina Laudis four decades ago, and has had the company of an African Grey for the past seventeen years.

I tried to read as many of this torrent of e-mails as possible, but often I could not, either because other demands took my time or because it was so very hard. Occasionally Arlene Levin-Rowe, my wonderful lab manager, got Alex's trainers and caregivers together in the lab for collective readings. It was always emotional. How could it not be? There we were in the smallish room at Brandeis that housed three cages: Griffin's, the one nearest the door; Wart's, to the right and the back; and the third cage, to the left and back of the room, with parrot toys scattered here and there, cage door open. Empty. This final e-mail I will share with you was one of those we read as a group. Its message prompted more tears than usual, for Alex and for the messenger whose heart he touched so dearly:

"I just wanted to write this and say that I have been suffering from a clinical depression for weeks, very numb inside despite being surrounded by a loving family and nearly two hundred beloved pets, my children, and grandbabies. Since I learned of Alex's death I have been reading the e-mails on the Web site, and tears are finally flowing. It is just another way Alex impacted the world. I forgot how to feel emotions, and reading Alex's nighttime words to his dearly beloved friend [i.e., "I love you"] opened the floodgates for me. Thank you, Alex, for touching my heart and helping me to feel again." The person who wrote these words was Deborah Younce, of Michigan.

Regular mail began to arrive as well, eventually boxes upon boxes upon boxes of it. One note was a beautiful card from Penny Patterson and friends, owner of the famous signing gorilla Koko. "Koko sends a message with the color of healing," Penny wrote. "Please know you are all in our thoughts and prayers—Alex's passing is a great loss to all." Below Penny's words was an orange squiggle, executed by Koko. Another note came from my friend and colleague Roger Fouts, longtime trainer and companion of Washoe, the famous signing chimp. "We know how you must feel," he said. "However, we are all getting old, and in our case we feel fortunate that Washoe has stayed with us as long as she has." Sadly, not too many weeks would go by before I sent Roger my condolences on Washoe's passing.

Treva Mathur of Trees for Life, in Wichita, Kansas, sent

me a certificate indicating that a gift of ten trees had been made to the foundation by the Windhover Veterinary Center, in Walpole, Massachusetts—a beautiful means for transforming kind thoughts into sustainable ecosystems. Alex had been an occasional (and reluctant) patient at Windhover.

But one of the most precious items that came in the mail was a package from Butler Elementary School in Lockport, Illinois. It contained about two dozen folders, each handmade by the children in Mrs. Karen Kraynak's fourth-grade class. On the front of each simple folder was the child's own delightful drawing of Alex, and inside a letter to me. Karen included a letter with the folders, explaining to me that she had become an African Grey owner after seeing the PBS film *Parrots: Look Who's Talking* about a decade ago. "Whenever I teach the children the bird section of my vertebrate course, I show them the PBS video and pictures of my own bird," she explained. "I happened to be doing this section when I read about Alex's death, and I talked about it in class. They knew how much my Grey means to me, and they understood how much Alex must have meant to you, Irene. We talked about what we could do, and the children decided they wanted to make sympathy cards." Here are just a few of the messages they sent me:

"I know Alex meant a lot to you," one of them began. "Inside of you will be OK in time."

Another letter started, "I feel sorry that your friend Alex left you. But he's in a better place now."

One was especially moving: "Alex must have meant a lot to you. He will always be with you. I lost my grandmother a few years ago. But deep down she is always with me. Just like Alex is always with you." The sentiments of children's innocent hearts touched ours so deeply, and moved us all to great tears.

On September 28, just three weeks after Alex died, I flew to Wichita, Kansas, and checked into the Hyatt Regency. I was there for a fund-raiser for The Alex Foundation that had been arranged several months earlier. There was to be a small gathering, a cocktail party for me to meet special donors, and then a larger number of people at dinner, when I was due to give a talk. Everyone there would be parrot enthusiasts.

I had given such talks dozens of times over the years, all over the country. I always presented Alex's most recent accomplishments, filled in some background about his other abilities to give the larger picture of his achievements, and then entertained questions. These seminars were always lively, always positive, always inspiring. I was always at ease at such events, never needing to think too much about what I would say. It was all so much a part of who I was. When I had set out from Boston on this occasion, I thought I would do my usual thing, the easiest path to take. By the time I landed in Wichita, I wasn't so sure. And when I turned in that night I knew it would be impossible, that I would have to do something different. After all, this was to be my first public talk since Alex passed away.

At the cocktail party, people spoke kindly and sympathetically to me; it was the same over dinner. The setting was Hyatt elegance, the excellent food beautifully presented. When it came time for my talk, I stood, looked around at all the faces turned in my direction, and thought, *What am I going to say?* I had no notes, even though the talk I was about to give would be completely new, different from anything I had ever given. I had decided I would just wing it and see what happened. I began to talk about the thousands of e-mails and letters we'd received, giving some examples of the sentiments people expressed. I told them of former students who wrote to tell me how working with Alex had influenced their career and life choices, how much they admired my strength in the face of enormous difficulties, how I always overcame the odds. And I told the intimate gathering that day that I had never seen myself as a strong woman, ever.

As I stood there talking, a kind of half-conscious sublimation occurred in the back of my mind, a growing crystallization of what the great outpouring of personal emotion and public recognition really meant. As this was happening, I could hear myself recounting people's stories about how Alex had impacted their lives, helped them in painful times. I read to them Wren Grahame's long e-mail about the miracle that was Alex and how this brought a miracle to her, and I spoke of how very moving it had been for me when I first read it. I spoke about the articles in the *New York Times* and other papers, the obituary in the *Economist*, the report in *Nature*, and all the other coverage that lauded Alex's (and my) accomplishments over the years.

It was an overwhelmingly emotional moment for me at the Hyatt that late September day. I didn't actually cry, but tears were ever present; I had to pause more than once. I could see tears at every table. Through all this welled up inside me the realization that had been growing over the previous few weeks—that what Alex and I had done in our time together had achieved important things in the world and in people's lives.

This realization was important because, over the years, despite Alex's accomplishments, he and I had faced a lot of denigration. You might think that an MIT- and Harvard-trained scientist working at various universities would be given a certain amount of deference, but as a woman working with a bird, I found it was sometimes the opposite. Some people argued that Alex was merely mimicking our voices, not thinking. Some said that my claims about animal minds were vacuous. This negativity had worn on me, had been like a weight on my self-confidence, my self-esteem. For thirty years I felt that I had been banging my head against a brick wall.

And now that weight seemed to lift. Stories such as Wren Grahame's and Deborah Younce's, and many, many others like them, touched me most deeply and made clear the impact Alex and I had had on people's lives. I had never been aware of it. I came to call this my *Wonderful Life* moment. In the 1946 film by that name, George Bailey (played by Jimmy Stewart), a small-time bank employee in middle America somewhere, was so depressed by what he saw as his unfulfilled life that he decided to commit suicide on Christmas Eve. Just as George is about to throw himself into an icy river, Clarence, an angel second class in need of earning his wings, stops him. Clarence takes George on flashbacks through his life, showing how his many small actions over the years had helped many, many people in ways in which George had been unaware. For me at that moment in Wichita, my Clarence was all the wonderful people and their messages, allowing me to see what had been there all along, invisible to me: Alex's and my work together had not been in vain.

That epiphany has allowed me to reconsider my own story, and Alex's, from the beginning.

Chapter 2

||||||||||||||||||||||||

Beginnings

My connection with birds goes back a long way. Not long after my fourth birthday my dad gave me a baby parakeet, a budgie, as a surprise gift. He was a little bundle of green feathers topped by a small head that darted nervously this way and that. I watched as the poor little thing trembled with anxiety, stepping rapidly from foot to foot on his perch, chirping tentatively: *peep, peep, peep.* Then he cocked his head, looked at me, cocked his head the other way, looked at me some more, and chirped with a little more confidence: *peep, peep, peep!* I was transfixed. "Hello, birdie," I said in my own tentative, four-year-old voice. I opened the cage door, offered my index finger, and he stepped on it. I lifted him up so we could see eye to eye and said, "Hello, little birdie. Who are you? What are we going to call you?"

"Let's call him Corkie," my dad said. Corkie had been my father's nickname as a kid, for what reason I never knew. "No," I said. "He's my bird, and I'm going to call him—"

I have tried very hard to recall what I said, but the name that so readily came to me that day has eluded me for some time now. That little bundle of green was to become so important in my early life, yet I've blocked out his name. We'll just have to call him No-Name for the purpose of telling my story. And that's not completely inappropriate, because in truth, at that point in my young life I felt a little bit like a no-name entity myself. You see, I was an only child, and there were no other children in our Brooklyn neighborhood. All my parents' friends lived quite far from us, and their children were a lot older than me. And my cousin Arlene, who was six months younger than me and therefore a potential friend, lived in Queens, a bit of a trek back then; visits were infrequent at best. So it was essentially just me.

My mom was what in those days would have been called a "refrigerator" parent: cold and distant, she never hugged me spontaneously or spoke loving words to me, and she never played with me nor read to me. My dad

taught elementary school during the day, studied thanks
to the GI Bill for a master's degree at night, and took care
of his sick mother, so I didn't see him from one "good
morning" kiss to the next. Until the day that No-Name
entered my life it had been just me, a loner with no one to
talk to all day but myself. But now that had changed. Now
it was no longer just me. Now it was No-Name and me. I
was thrilled. I had a companion at last, someone to talk to,
someone who appeared to be devoted to me.

My parents' Brooklyn house was on Utica Avenue, not
far from Eastern Parkway. It was about as urban a setting
as could be imagined. We lived in the second-floor apartment of a two-story turn-of-the-century redbrick building
that my dad had inherited from his father. The stairs that
led up to our apartment looked to my young eyes to be
precipitous and endless. The ground floor was a store that
was rented to a succession of businesses, whoever could
pay the rent. In back was a small guest house where my
uncle Harold lived. I hardly ever saw him.

Our apartment was quite large, with two bedrooms
in the front overlooking the street: one was my parents',
while the other was a guest room, though no guest ever
stayed there as far as I knew. A central living room was

dominated by my father's proudest possession, a Victrola phonograph, a monster piece of equipment, all shiny wooden veneer, speakers, and brass handles. I spent a lot of my time dancing by myself to Strauss waltzes I played on that Victrola, spinning and twirling, whirling and swooshing. How I got into such a pastime at that age, and what I was thinking as I danced, I do not remember. But I do remember the sense of freedom and joy of moving to the music.

My room was at the back, overlooking the yard, such as it was. I especially liked the circus wallpaper: elephants, tents, and clowns. At the back also was my father's "workroom," where in the little spare time he had he sculpted clay figures, usually human heads. I loved to watch him as a nose, lips, and ears emerged from the lifeless clay as if by magic. A door from the workroom led onto a spacious, open porch edged by low cinder-block walls, plastered and whitewashed. In the summer I had one of those little inflatable kiddie pools out there, where I played by myself. And all around were flowers in boxes and pots, ablaze with color. My father spent hours tending to them in solitary absorption, as with his sculpting. When we later moved to a house, this absorption would turn to growing African violets in the basement in winter and to a wondrous real garden in summer.

Beginnings

I spent a lot of my time watching kids' shows on television in the morning. Then I would draw, developing some little talent I inherited from my father. Or I'd fill in coloring books my aunt gave me. My mom and dad disapproved of spending money for such things—or maybe didn't have any to spare—and instead my dad drew circles and other shapes on pieces of paper, so I could fill them in like Easter eggs. I was never given toys, mostly because my parents had never been given toys in their childhoods, and so it was something they didn't even think about. Both my parents were first-generation Americans: my mom's parents were Romanian, my dad's Lithuanian, and both had experienced severe deprivation growing up. In any case, the toys didn't matter to me. I was just as happy playing with pots and pans, taking the coffeepot apart and putting it back together. But buttons were my favorite.

My mom had a huge button drawer. Her father had been in the rag trade, and so there was an endless supply of buttons of all kinds. I spent hours playing with them, sorting them into categories, some obvious, such as size, color, and shape, and some I made up. Sometimes I sat at the coffee table, meticulously organizing, carefully sorting. Other times I lay on the floor in my bedroom, buttons at eye level, seeing close up this ever-moving kaleidoscope of my own making, often seeming to develop a life of its own. Of course, I had to do all this game playing, what-ever it was, without "making a mess," as my mother con-stantly admonished me.

When No-Name came into my life he fell right into my daily routine. He sat on my shoulder as I watched televi-sion or colored, and chirped a lot. But his favorite, like mine, was the button drawer. As I did my sorting routine, No-Name darted about from pile to pile, pushing buttons this way and that. It became quite a game to see if my attempts at imposing order could keep ahead of his high-energy disruptions.

We also loved the typewriter game. My dad had a typewriter in his workroom, the manual kind with the carriage you flipped back to the beginning of the line by pushing a lever, which produced a cheerful *bing* when a hammer struck a bell. In the days before No-Name, I often spent what seemed like hours banging away at the keyboard, sweeping the carriage back—*bing*—over and over. Now, with No-Name as part of the game, it was much more fun. He sat on the carriage as I banged the keys, jerking unsteadily along with each hammer stroke. And he seemed to love it when I swept the carriage back and produced the *bing*. He chirped cheerily at that and hopped around happily.

•   •   •

No-Name never learned words. He never spoke to me. He didn't need to, as far as I was concerned. I talked to him endlessly, about anything and everything, and he would look at me intently and chirp enthusiastically right back. To a four-and-a-half-year-old who had yearned for companionship, ached for love, No-Name offered a lot. In her book *Animal Dreams*, Barbara Kingsolver wrote, "Children robbed of love will dwell on magic." For me back then, I could imagine no greater magic coming into my life than having the intimacy and love I felt with my little bundle of green feathers, No-Name.

My mom was bitter about her life, with good reason. When she was first married she had a good job as a bookkeeper at a housing project, which she loved. She expected to go on to better things. Then she became pregnant with me, and it meant the end of her professional life. This was 1948, and mothers didn't work back then. She had no choice but to give up her job, her hope of a fulfilling future gone. She bitterly resented it. Mom made it very clear to me—often explicitly—that I was the reason for her ruined life, her consignment to a permanent sentence of drudgery. She occupied her days obsessively washing, ironing, cleaning the apartment, cooking (even though she didn't like to eat), and then cleaning all over again. Most days she would go to the local shops. I usually tagged along with her.

Once, not long before No-Name came into my life, we went into the local bakery to buy freshly baked bread, a treat, because we usually had the homogenized white stuff. (I remember this incident because it was so painful and bewildering to me at the time. It is also iconic of my life back then.) The woman behind the counter in the shop picked up a cookie and offered it to me with a big smile. "Here," she said to me, "would you like a cookie, little girl?" I was extremely shy with everyone, simply because I had had so little social interaction with anyone at all, especially strangers, and was completely socially inept, even for a four-year-old. Head bent, staring intently at the floor, probably hoping it would swallow me up and thus enable me to avoid having to deal with the situation, I stretched out my hand and took the cookie. Didn't say a word.

The woman must have been a little taken aback by my mute, gauche behavior. She said, "What do you say, little girl?" I had no idea what she was talking about, because my mom had not taught me what to say in these (or any) circumstances. Nothing about "please" and "thank you." I guess Mom thought I would just pick it up. So I just stood there, head still bent in painful shyness, still not saying anything. "Well, you'll just have to give it back, won't you," the woman said, probably teasing me. So I handed it back, desperately trying not to cry.

My mom was mortified. She thought of herself as being "quite proper." She made embarrassed excuses about my awkwardness and shyness, and pushed me through the doorway and into the street. Then she berated me all the way home for embarrassing her in front of the baker's wife. I had no idea what she was talking about. All I knew was that I had behaved badly somehow, failed her somehow.

Imagine taking this same untutored, socially inept loner of a kid—me—and thrusting her into the local school at the age of five. Imagine then that she is the only white child in the class. Kids made fun of me endlessly. I had funny hair. I had funny skin. It was traumatic enough to switch so abruptly from social isolation in my parents' quiet apartment to immersion among thirty or so raucous kids in a small classroom. The taunting on top of that was torture. The kids weren't being intentionally cruel, I'm sure, just kids being kids. But the effect was the same. Pretty soon I started getting sick a lot, all kinds of symptoms, missing school, obviously suffering from something. After the pediatrician could find nothing really wrong, my dad took me to a child psychologist, who essentially said, "The school is a toxic environment for her. Get her out of there."

Within six months we had moved from Brooklyn to Mentone Avenue in Laurelton, Queens, not far from my aunt (my mom's sister), uncle, and my cousin Arlene. Their house was north of Merrick Road, ours was south, in a less favorable part of town where the houses abutted the Long Island Rail Road tracks. Our neighborhood was simple, squarish houses, separated by little driveways, with decent-sized yards in the back. They were built just after the war for returning servicemen. The effect was homogenous and nondescript but pleasantly leafy, certainly in comparison with our Brooklyn neighborhood.

Our yard had a big mulberry tree in the back, which attracted birds all summer, and I thrilled to that, given my newfound love of birds. Dad put up a bird feeder so we could extend bird-watching throughout most of the year. The backyard faced those railroad tracks, and every time a train passed by, the house shook a little. I came to find the noise and the vibration a comfort, in the way that familiar elements in one's environment can be. Arlene, by contrast, when she visited, was in constant fear that the train would roll right down into the yard and kill us all.

The move was good for my dad because he could now really indulge his passion for flowers. He planted them profusely. The garden was his joy, his love. Nothing much changed for my mom, however, just different surroundings in which to mourn her lot. She and my dad were fighting (verbally) more and more, and I frequently retreated to the attic to escape their harsh words. I used to go there to read and do my artwork, too.

My parents were dealing with their own demons, of course, but as a child I was unaware of them; all I felt were their consequences. When my mom was sixteen, her mother died, and the job of cooking and housekeeping for herself, her three siblings, and her father, fell to her. Her father did allow my mom to finish the last couple of months of high school before taking on what must have been an enormous burden for a young girl. As an adult, she must have yearned to be taken care of, not to be thrust into caregiving once again. And her constant fearfulness—fear of anything new, fear of getting lost in the car when my dad was driving, fear of driving herself—were all probably rooted in the grinding uncertainty of her Depression-era childhood and then the terror of her new husband being in the war, with no word for months on end. She was attractive, very photogenic, always elegant when outside our home, always waiting for what would never be.

As for my dad's volcanic temper and obsessive need for control, I eventually came to understand it as part of the trauma of the war. He talked occasionally about the war to me, but only in the vaguest of terms, usually making light of it. If I tried to get more detail, he would either change the subject or make a joke, often about the ineptitude of his commanders, *M*A*S*H*-style. He obviously didn't want to talk about the horrors. I learned only recently that he had been in the Battle of the Bulge, lived through scenes of horrendous carnage and weeks of deprivation, and been badly injured himself, physically but mostly psychologically. Silent struggles with the past, for both of them.

My own struggles for companionship took a turn for the better in Queens. I owned a succession of parakeets following the death of No-Name, each of whom I still remember. There were Greeny and Bluey, several Charlie Birds, and more. Dime-store birds, they didn't last very long, as no one understood proper diet and no one thought to take a sick bird that cost a few dollars to the veterinarian. Charlie Bird number one was the first real talker. Considering that I remember him and the others, it is all the odder to me that I cannot fill in the identity of No-Name.

As far as human companionship was concerned, things improved only slowly. I was a thoroughgoing nerd, at one point complete with a pair of Coke-bottle-thick, blue-rimmed cat's-eye glasses. I freely admit it, and I have school photos to prove it. I was even accelerated a couple of grades, so I was again rather socially inept, surrounded as I was by older kids.

I had just two guests at my first birthday party in Queens—my cousin Arlene and the man who was painting the decorative disaster of a house we had moved into. I think people were initially wary about who would want to buy such a run-down mess, and so they steered clear of us. But before too long I began to make a few friends. I thus discovered that I wasn't the inherent loner or social recluse I had grown up imagining myself to be. Not that I was much of a socialite, either.

During the summer most of the kids in the neighborhood went to some form of camp. I stayed home, rode my bike around the neighborhood, and read endlessly, a passion I shared with my dad. (My cousin Arlene recently reminded me that we even read books at the dinner table when she visited.) It's no surprise that when I eventually read *Dr. Doolittle*, I was completely enthralled. Here I was, talking back and forth with Charlie Bird number one. And here was this character who learned to talk to animals and understood what animals were saying. Dr. Doolittle is taught to converse with animals initially by an African Grey parrot named Polynesia. I daydreamed a lot about that, about being able to talk to and understand animals' speech and thoughts.

. . .

I had one close friend at high school, Doris Wiener, and a small group of other friends, mostly guys. Our connection principally was that we were all very studious. Most of the group were somewhat unfocused, but Doris and I were on a serious science track, the lone females in many of our classes. This was in the sixties, remember, so we were considered to be rather odd, and certainly not desirable as girls. We were also judged to be bright, so we were part of a group of fifty or so out of a graduating class of over a thousand who all took the same advanced classes.

So in my final couple of years at high school I was a certified nerd and two years younger than many of my classmates; I was fourteen at the beginning of my junior year. The girls were mostly wearing the exaggerated makeup of the time and parading around in stylish clothes. I wore almost no makeup (a tiny smudge of light brown eyeliner at most) and had to make do with hand-me-down clothes. Despite all this, I did develop a certain confidence in who I was, partly through a flowering and nurturing of a passion for classical music (remember those Strauss waltzes) and the theater: we used to get highly discounted tickets to Broadway shows and Carnegie Hall as part of our advanced classes, and I loved every minute of those excursions. And I began to find out who I was intellectually. I was extremely analytical, and good at it.

I began to realize this analytical part of me when I was first introduced to the periodic table of elements in chemistry class. Our task was to learn it, a giant memorization challenge, with ninety-plus elements arrayed in rows and columns. Then we were to learn how the elements reacted with one another. I was blessed with something of a photographic memory, so learning stuff was not a problem: it's why I did so well at history and French, for instance. I started memorizing the information about the elements, and it quickly dawned on me that there were patterns and a pervasive order that made it all predictable.

I discovered that once you knew how sodium reacts, for instance, you also knew how potassium reacts. Once you knew where any element was on the table, you pretty much knew how it would react with other elements. I adored it, the predictability of the patterns. It wasn't about memorizing a bunch of garbage; it was about logic, and I was seduced by what I felt was its compelling beauty. I was very good at French, too, and won prizes for that. But I knew I was going to have to support myself at some point, and science seemed to offer better prospects than the humanities. Throughout much of high school I had assumed I would pursue some kind of career in the biological sciences. My dad definitely encouraged me in that direction because of his own interests; he had hoped to be a biochemist, but the Depression and World War II intervened. After the epiphany of the periodic table, I knew it would be chemistry, not biology. I was hooked.

I was so hooked that I elected to be part of a group of twenty-four other students from high schools around Queens, almost all guys, to do a crash course of freshman college chemistry the summer before my senior year. It was a whole year's worth of college chemistry in just a bit over six weeks of the summer break, instead of going to the beach or some other "normal" teenage summer activity. I thought it would be fun, but in fact the trek to Queens College, in Flushing, was wearing. And that paled in the face of the unrelenting intensity of the course itself.

I did well enough in the course, but I can readily admit that it was absolutely horrible. What had I been thinking? There was one high point, however, toward the end of the six weeks. We were in lab class, and the lecturer was instructing us as to what we were supposed to be doing. He had a lab assistant who was all too obviously not happy to be there with us that summer. He was probably a graduate student just cursing his ill luck that he had to take care of these high school kids and keep them from killing themselves in the lab. The windows were open, because it was hot, and in flew a yellow parakeet, clearly distressed, flying back and forth among the lab benches. Bunsen burners were going, and there was all this dangerous lab equipment for the parakeet to negotiate. The lecturer was shrieking, "Get it out! Get it out!"

I shouted, "No, no, it's OK. I'll get it."

I yelled for everyone to turn off their Bunsen burners, I put out a saucer of water in the corner of the room, and I asked everyone to be calm and quiet, so as not to freak out the bird any more than it already was. Very soon the bird landed on the saucer and started drinking vigorously. The poor thing was obviously very thirsty. I was able to catch it, and took it home. I planned to keep it, but Charlie Bird had other ideas. He started to fight with the unwelcome (to him) newcomer, and I realized I had to put an ad in the paper to try to locate its owner. The next day a girl called on the phone and said, crying, "I know this isn't my bird, but I just lost my bird, and I am really sad. If nobody else wants this bird, I really want this bird." I (and probably Charlie Bird) happily surrendered it to her.

In whimsical moments, I look back at this little story and think that it was as if the universe were trying to remind me of where my heart was: biology, and particularly birds, not chemistry. But my mind was set. The only question was, where to go to college?

. . .

I aimed high, and why not? I graduated third in a class of sixteen hundred students, and I was only sixteen. I initially considered Cornell, a very good Ivy League school, because other girls had gone there from my school, as had a cousin. But two things happened. First, my parents learned that I could apply to the Ag School, the state-supported land-grant part, and insisted that I do so; it would save them lots of money but make majoring in chemistry rather difficult. Second, I discovered that the town was essentially fourteen bars and two movie theaters. I nixed that option. I am being a little facetious here, of course, but I definitely wanted to be in a town with a strong arts presence. I visited Boston and immediately fell in love with the city, because it had all the theater and music I could possibly need. The obvious choice was Radcliffe, which meant I'd study at Harvard, which had a very good chemistry department. When I told my college counselor of my choice, he said, "Why not try for Vassar?"

My response was, "Vassar, that's a girls' school. Why would I want to go there? I'm a *chemistry* major."

Then he surprised me and said, "OK, so why don't you apply to MIT?"

I was shocked. "What? Girls don't go to MIT."

"Yes, they do," he said. "A couple of them, anyway." One of them was from my high school and was now in graduate school. She was going to be visiting over the holidays, so he arranged for us to meet.

"Yeah, there are some girls there," she told me. "MIT is actually trying to increase the number of female students. Yeah, there's, like, twenty, thirty girls each year." I figured, hey, it was right next to Boston, why not give it a try?

Radcliffe put me on a wait list. MIT accepted me, so, five months after my sixteenth birthday, with much still to understand about how to conduct a social life, not to mention how to live away from home, I packed my bags and stepped into this daunting, machismo bastion of revered learning. Charlie Bird number two came with me, though not till my sophomore year, when I had a room to myself.

Charlie Bird was my constant companion—and solace—in the high-pressure academic environment that was (and still is) MIT. Trying to cope with the onslaught of course work that is part of the accepted culture there is often likened to trying to drink from a fire hose. That, combined with the hypernerdiness that is also part of the culture, can make for a pretty lonely and miserable experience for a student, especially for a not very socially sophisticated girl. When I got back to my dorm room each evening, Charlie Bird always gave me a warm, chirrupy greeting that was very welcome relief after the grind of the day. He sat with me as I tackled the assigned reading each evening, preening his pretty green plumage, singing and talking. Our "conversation" was often my only non-work-related exchange of the whole day, at least for the early part of my four years there.

One time, when I went to a teaching assistant to ask some follow-up questions from a meeting we'd had a week earlier, he said to me: "I know this is a very weird question, but after you left last week, there were all these little green feathers on the floor. What was that about?" Of course, they were Charlie Bird's feathers that had got caught in the book as he preened and perched on the spine while I was reading and turning the pages. They had drifted onto the floor and in between the pages as the assistant and I leafed through the book. This is one of the few memories of my early years at MIT that always brings a smile to my face, even now.

Money was always tight for me, not least because of the horrendous cost of tuition, housing, and books, despite an MIT scholarship and the little help my parents were able to give me. I tried to economize the last two years by living essentially on tomato juice, boiled eggs, instant coffee, and ice cream, the last from a small café on campus. The ice cream guy quickly realized what I was up to, and he started to give me extra scoops for free.

The hardships and social awkwardness aside, I became ever more enthralled with chemistry, especially theoretical chemistry, and the pattern, order, and predictability of its equations. I also became enthralled with a man. David Pepperberg, a graduate student at MIT, was having trouble coming to grips with organic chemistry, then my forte, and I was having trouble with electricity and magnetism, his forte. So we tutored each other. Before very long we were going steady.

At this point I imagined myself pursing a career in chemistry, probably as a university professor. This was still something of a disappointment to my dad because of his love of biology. But, he said, at least it was a real science. Graduate school was therefore an essential next step for me, and, as David had yet to finish his Ph.D. thesis, I didn't want to go very far away from Cambridge. When I applied to Harvard to do theoretical chemistry, my friends told me, "You're nuts to even think about it." Harvard's chemistry department had a world-class academic reputation. It also had a world-class reputation for being excessively male-oriented. Women were seldom seen there. (It also, I learned much later, had an excessively high suicide rate for academic departments of its kind, and having survived the grueling, pressure-cooker culture of the place, I am not surprised.)

As it happened, the year I applied, 1969, was the first year the United States government refused to give men a military draft deferment for going to graduate school. The Vietnam war had initially boosted enrollments; now the draft suppressed them. The department was forced to take in many more women than its usual token one per class, because it needed teaching assistants. I was one of some half-dozen women in a class of fifteen. I very quickly got a clear view of how women were viewed in the masculine-oriented world of advanced chemistry.

David and I became engaged soon after I started at Harvard, and I proudly wore an antique ring with a big diamond; it had belonged to David's grandmother. Just before Easter, I went into the administration office for some formality to do with my course. "Oh, is that an engagement ring?" the woman behind the administrator's desk asked me cheerily. I said it was, and stretched out my hand to show it to her proudly. She then said, "So, when are you leaving?"

The holiday was coming up, so I said, "We're leaving for Easter break a bit early, on Wednesday afternoon."

The woman looked bewildered, shook her head, and said, "No, no, no. I mean, you know, when are you withdrawing from the department?"

"Why would I do that?" I said, having no idea what she was talking about.

She pointed to my ring and said, "Because you're engaged," as if that were all the explanation needed. She obviously thought I, as a married woman, should stay home, keep house for my husband, and produce babies; at most, take the kind of undemanding job that she had, and certainly not occupy valuable space in her department, thus excluding a man from that rightful honor.

I told her I had no intention of leaving, and walked out of the room. I wasn't going to be forced to do what my mom had had to do years earlier.

After David and I married, he moved into my tiny studio apartment on the middle floor of a three-story house on Hammond Street, just behind Harvard's Divinity School, one of Cambridge's cozier neighborhoods. Charlie Bird came, too. It was not an easy life, with David doing experiments that might run for thirty-six hours at a time, coming home at odd hours, and me taking difficult courses and trying to develop a research program.

After a few years had passed, my once-burning love affair with theoretical chemistry began to cool. Part of the disillusionment was my changing perception of career prospects. I heard from women in my cohort who were doing nontheoretical studies and thus nearing graduation that they were facing strong anti-female bias in job recruiting. They were asked questions like "What kind of birth control are you using?" and "So, you're married; when are you going to have children and leave?" This was the early seventies, and the feminist movement still had a long way to go.

Also eroding my passion was the subject itself. I wanted to figure out how molecules might interact, how reactions might happen, based on an understanding of their fundamental properties. Instead, I was spending more and more of my time running programs on IBM mainframe computers, long, complicated calculations involving endless typing of punch cards and then more hours of finding the one stupid little typo that crashed the program. Computers were still primitive, and working with them was laborious and dull. I was ready for a change, but I didn't quite realize it. I needed a push.

Beginnings

. . .

A celebrated local pyromaniac provided the shove. During the night of November 8, 1973, he set fire to garages abutting each of five houses in Cambridge. Our Hammond Street house was the last on his list, so by that time the town's fire teams were pretty stretched, and we had to wait for a truck from neighboring Somerville. The house was destroyed, and we escaped with nothing but the clothes on our backs. Luckily, David had turned in his thesis two weeks earlier, and Chet, our last parakeet, had passed away a week previously, a victim, we think, of carbon monoxide drifting up from the garage on the ground floor. We were now homeless and without possessions.

Harvard took pity on me and waived tuition fees for a semester. John Dowling, David's postdoctoral adviser, took us in at his house in Lincoln, about ten miles west of Cambridge; in return, I cooked dinners and we helped look after his two young sons. It was a traumatic, tumultuous time, dealing with the dislocation and loss. The following March, PBS debuted its *NOVA* television series, devoted to science and nature. In my previous life I would not have seen these programs, because we didn't have much time to watch television. But because we were in John's house, we sometimes did, particularly if a program was on that was somewhat educational and that the two boys might find interesting.

Among the early programs we saw were reports on dolphins whistling and chimps signing under the tutelage of university researchers. A later one was on why birds sing. I still remember the visceral shock of these shows. They were a revelation. Humans communicating with animals, animals communicating with humans, and humans learning about how animals learned to communicate with each other—it seemed little short of a miracle to me.

I had been vaguely aware of a woman called Jane Goodall who was studying chimps somewhere in Africa. I had also been vaguely aware that three European researchers—Karl von Frisch, Konrad Lorenz, and Nikolaas Tinbergen—had won the Nobel Prize the previous fall for their studies in some aspect of animal behavior, but I hadn't absorbed what it was or why it was important. I had no idea that serious researchers were doing serious studies on how animals lived in their natural state and on what was going on in their minds. And I definitely was not aware that a man called Donald Griffin, who had made his name by discovering how bats navigate (with sonar), was leading a revolution in the way biologists were beginning to think about animal minds, about animal thinking.

Remember, MIT wasn't exactly the kind of place that dealt with such topics, and I had had no exposure to this type of research.

But I did know—instantly and inescapably—that this was where my future lay. I had no idea what I would do nor how I would do it, but I recognized the moment as one of those rare "feelings" one sometimes gets in life that one just "knows" is right, that one just has to follow. Given my lack of exposure to whole-animal biology—high school classes were about the digestive system and such—it was not really surprising that, until that instant, I had given not a single, serious thought to studying animals as a career; I had not once lain awake at night thinking, *Gee, I wish I was able to study human-animal communication rather than slogging on with chemistry that I am no longer enjoying.* But here I was, prepared to throw aside years of university work, of deep commitment and effort to studying chemistry as a career path, and embark on a venture for which I had little knowledge and no training.

Our host, John Dowling, was a professor in Harvard's biology department, so he was in an excellent position to orient me. He essentially said to me: "Yes, studying animal behavior is real science, and we do some of that here at Harvard. If you are seriously interested in pursuing this, why don't you go to the Museum of Comparative

Zoology and talk to people there?" I did, and as a result of what I was told, I started attending courses and seminars in bird behavior, child cognition, and language. And I read voraciously: anything and everything I needed to equip me for where I wanted to go. I continued to put in the hours necessary to finish my doctorate in chemistry, but I had a new calling.

I learned about the pioneering work on human-chimp communication by people such as Allen and Beatrice Gardner, David Premack, and Duane Rumbaugh. I heard Peter Marler talk about his discoveries on how birds learn their songs. I was enraptured by this whole new area of science: new, definitely, to me, but new also to science itself, inasmuch as these people were breaking new ground. It was hallowed ground. They were talking about teaching nonhuman animals the rudiments of human language and probing the extent of animal thinking and communication. Received scientific wisdom at the time insisted that animals were little more than robotic automatons, mindlessly responding to stimuli in their environment. The newly emerging science was overturning that view completely. It was nothing less than a revolution. And I wanted to be part of it.

My only question was, "What animal should I study?"

The answer was obvious. Birds learn their songs, and

Beginnings

I knew from my own experience with my parakeets that they can learn words (some of them, anyway). Others were working on human-animal communication using chimpanzees. No one was working with birds. I knew birds are smart, and I was confident they could do this.

Besides, from a practical point of view, working with birds is a lot easier than working with chimpanzees. I needed a species that could learn speech, which meant either parrots and related species or corvids (crows, ravens, and such). It didn't take much time to discover that parrots are better talkers than crows and their relatives, and that the species that learned most easily and was the clearest talker was the African Grey parrot. A Grey it would be.

Greys are now one of the most popular of bird pets. Indeed, parrots have a long history as pets, going back four thousand years. Egyptian hieroglyphics show images of pet parrots, and noble Greek and Roman families kept Greys, too. So did Henry VIII of England. And, of course, for a long time they were popular with Portuguese sailors, vocal companions on long voyages. They also happen to be exceedingly beautiful creatures, with delicate gray and white plumage, a white area around the eye, and a bright crimson tail. I also learned that Greys love attention and form deep and lasting bonds, with their owners becoming profoundly emotional about their birds.

I determined that this would not be true of my Grey and me, however: I chose the species as a study animal because they had been shown to be very smart; I wasn't choosing a pet. A German zoologist, Otto Koehler, had done groundbreaking work in the 1950s, showing that Greys have an unusual facility with numbers, and one of Koehler's assistants, Dietmar Todt, had shown that Greys readily learn speech through social interaction. Aside from that, not much else was known about them in the realm of science. But that was enough for me.

I finished my theoretical chemistry doctorate in May 1976, and David accepted a post in the biological sciences department at Purdue University, in West Lafayette, Indiana, beginning January 1, 1977. I hoped to find a way of starting my own avian research there. In June 1977, we drove to Noah's Ark, a pet store near O'Hare Airport in Chicago, to pick out my own Grey parrot. I had been in touch with the bird department director of Noah's Ark several times in the previous few months, and I knew that he had about eight birds that had been bred in captivity.

The place was huge, a cacophony of all kinds of potential pets and prospective owners. The bird director greeted us and showed us where the Greys were, a big cage with eight birds, all about a year old. "Which one would you like?" he said, looking at me.

I shrugged, because I didn't know how to choose. In any case, I reasoned that because I was embarking on a scientific study that should reflect the cognitive abilities of Greys in general, I thought it best to have one chosen at random. "Why don't you select one for me?" I said.

"OK," he replied, and picked up a net, opened the cage door, and scooped up the most convenient bird he could reach. He flipped the bird on its back on a table, clipped its wings, claws, and beak, and popped it into a small box. Very unceremonious.

The ride back to Lafayette was three and a half hours. It must have been hard for the little creature confined in the dark, having just been plucked away from the flock he had been with for at least half a year. I carried the traveling box to the lab space I had borrowed in the biological sciences department, and placed it on a table next to a proper parrot cage I had positioned in the corner of the room, so as to give him the best sense of security I could provide. David put on some heavy gloves, opened the box, picked up the struggling bird, and eventually managed to put him in the cage. (I always had David do the things that might traumatize the bird, because I needed to establish a sense of trust with it.)

Right at that moment, the bird obviously didn't trust anything or anyone, least of all me. He was trembling, squawking nervously, and stepping from foot to foot on his perch. The poor creature was obviously in a state of shock. He was also clearly scared of a parakeet, Merlin, in a cage at the other side of the room. Merlin was just as clearly scared of him.

My Grey hardly looked the part at that moment, trembling and insignificant looking as he was, but here was the bird I hoped—and expected—would come to change the way people would think about the minds of creatures other than ourselves. Here was the bird that was going to change my life forever. I couldn't help thinking back to No-Name, the parakeet who had transformed my life twenty-four years earlier. He had weighed barely an ounce and measured just a couple of inches long. My newly acquired Grey was much bigger, almost a pound in weight, and ten inches tall. But my Grey was just as nervous, just as fearful, as No-Name had been all that time ago.

This time, however, my new bird had a name. He was Alex.

Chapter 3

||||||||||||||||||||

Alex's First Labels

I'm not sure who was more nervous in our first days together, Alex or me. I know I was a little on edge, and he sure looked it, the poor traumatized bird. He'd been snatched from what had been his home for many months and thrust into a completely new environment, a small, fairly bare room occupied by a scary parakeet and unfamiliar humans. I considered myself a bird person, but I'd never had such a big bird before, and I was more than a little unsure about how best to handle him. I knew what food and drink to give him. I knew I needed to talk softly and soothingly to him at first, and give him treats. I understood that I had to build his trust in me.

It didn't start well. Alex was still uneasy on the second

day, still scared of the parakeet. I decided to move Merlin's cage to another room. I then went back to Alex and tried to encourage him to perch on my arm. He wouldn't even come out of the cage, despite my gentle verbal entreaties. The phone in the adjacent room rang; I went to answer it. By the time I returned to the lab, a minute at most, Alex had climbed out of his cage. *Yes! Progress.* I offered him some fruit, which he fussed with but didn't eat. I held out my arm for him to perch, and he clumsily climbed onto it. I imagined he had never perched on someone's arm before. *More progress.*

Not for long. Clearly still alarmed, Alex tried to fly, and promptly crashed to the floor because his wings had been clipped back at the pet store. He was squawking pathetically, flapping his wings wildly. Suddenly there was blood everywhere, spraying this way and that. He had broken a new wing feather. Poor Alex was freaking out, and so was I, but I tried to appear calm so as not to upset him any more than he was already. Having dealt with broken feathers with my parakeets, I knew what to do. But I was facing a very frightened and significantly larger bird here, not a comfortably established pet parakeet. That made it much harder, more hazardous. I eventually managed to gather him up, remove the feather, and get him back into his cage. He was obviously badly shaken. "Alex does not come out more that day, scared of me," I wrote in the journal I started when Alex arrived. Who could blame him?

Over the next few days Alex became a little braver, bit by tiny bit. He started to come out of his cage spontaneously, but was still very wary of me. On the third day he did perch on my hand, by accident: he had tried to avoid me, but found himself perching for a few seconds. I started to give him objects, such as paper and pieces of wood, to explore his preferences. I planned to begin by teaching him labels for things he liked, figuring it would speed up the learning process. It turned out that he loved paper index cards even more than food. He chewed them enthusiastically, rapidly tearing them to shreds.

Day four was even better. Alex again came out of his cage spontaneously and even perched voluntarily for a short time. He continued to enjoy chewing paper. When I gave him some I said things like, "*Paper*, here's your *paper*," placing emphasis on the relevant label. My friend Marion Pak, who volunteered to help train Alex, came to meet him for the first time. He immediately took to her, perched easily, and spent an hour seeming quite content with her. And why not? She wasn't the one who had subjected him to torture in a dark box for hours, tossed him on the floor, and broken a feather.

I needed Marion's help with Alex because I was

going to use a modified form of a training method I had researched while at Harvard. I'll describe it in more detail later. Essentially, though, the method involves two trainers, rather than the usual one, and they take turns asking each other about an object's label, with Alex observing. Then either one would query Alex, using the same words. The idea was that he would learn in a social context. This procedure was radically different from what would have been considered normal at the time. Marion and I started such training that day, on the label "paper."

After Marion left that morning I stayed with Alex for another hour. I purposely ignored him until he made a noise, then I rewarded him with paper, again saying, "*Paper*, Alex, here's your *paper*." Any parrot owner can tell you that his or her bird may spontaneously learn some random words, but that's not the same as teaching meaningful communication. The first small step in Alex's training was for me to link any novel sound to the single object, paper, as Marion and I had done in training earlier. The only vocalization Alex made was something like "Auf," which seemed exploratory, and a rasping, subvocal noise he made randomly. As I gave him one index card I said, "OK, Alex, there's a long way to go, buddy." Alex didn't say anything, just continued shredding paper and wiping his beak. But we had started our work together at last.

It turned out that beginning training with "paper" was an extremely bad choice, because it is very hard to make a "puh" sound if you don't have lips. But Alex himself had made the choice, so we were stuck with it.

During the next four or five weeks I steadily raised the bar for Alex, to push him to achieve more and more. For instance, during training, Marion and I waited for some kind of two-syllable utterance—resembling "pa-per" in rhythm if not in actual sounds—before we would reward him with some paper. That's what is called the "acoustic envelope," the sound shape of the word. We also introduced a silver-colored key to Alex, so that he wouldn't come to associate verbalizing only with paper. He became steadily more vocal and began to produce sounds like "ay-er" when Marion and I asked him, "What's this?" when showing him paper, and "ee" for the key. Sometimes he got confused and combined the sounds, like "ee-er." But he was definitely beginning to get it.

Within just a few weeks of beginning training, Alex was indisputably using vocal labels to identify specific objects. He was not merely mimicking, or parroting, us. The first real indication of this happened on July 1. I had seen that he liked to use paper to clean his beak, especially after eating something messy, such as fruit. I'd often give him apple so he'd need paper, which initially he would indicate by some fairly indecipherable vocalizing. That day, however, I gave him apple but forgot about the paper. He was on top of his cage, as usual, and gave me one of those *OK, what's the problem, lady?* looks that he would hone to perfection over the years. He ambled to the edge of the cage, looked down to where I stored the index cards in a drawer, and said "ay-ah," or something very like that. It certainly wasn't his spontaneous little gravely sound.

I was thrilled, but wanted to make sure it wasn't accidental. I gave him paper to reward that first "ay-ah," and he chewed it happily for a while. I then held up another piece and asked him what it was. He said "ay-ah" again. And I rewarded him again. That happened half a dozen times. The seventh time he'd obviously had enough. He began preening energetically and occasionally chattering in his gravely voice. Alex was always capable of letting me know when he was tired of lessons!

"What a day!" That's how I began my journal entry on August 4. Marion trained with me again that day. "[Alex] did amazingly well!" I wrote. "He corrected himself, gave us the objects—even improved on pronunciation." He'd produced the best "puh" sound so far, saying "pay-er." And his accuracy at labeling "key" jumped dramatically. "It was as if he finally made the connection," I wrote triumphantly. It was a "I think he's got it; by George, he's got it" moment.

The next day's journal entry starts: "Alex incredibly stupid today! He acts as tho' he's forgotten yesterday existed! Almost impossible to get him to say a decent KEY. PAPER never clear. What happened?" It was very frustrating, to say the least. I was bewildered. For his part, Alex seemed quite content. He happily ate banana when I gave it to him, and made soft sounds. And he was beginning to look resplendent, as new feathers were coming in to replace ones he'd plucked early on, in his initial nervousness. I especially admired the new crimson tail feathers. But "key" and "paper" appeared to be far from his mind.

Only later did we learn that this pattern of behavior is quite normal. The Swiss psychologist Jean Piaget has argued that when children learn something new, they need time to assimilate it before using it with ease. When we started to tape Alex's solitary evening babbling some years later, we frequently heard him "practice" a newly acquired word very clearly, even though earlier in the day he had completely failed to say it. Quite likely, during those evenings of August 4 and 5, Alex was cheerfully producing lucid renditions of "pay-er" and "key" to himself repeatedly. But we had no way of knowing that.

A little later, he displayed another clue to his increasing understanding of sounds as labels. A few weeks after the "by George, he's got it" moment, he correctly identified a red key as "key," even though we had trained him only on silver keys: he knew that a key was a key, whatever its color. It was his first demonstration of what psychologists call "transfer." This kind of *vocal* cognitive ability had never before been demonstrated in nonhuman animals, not even in chimpanzees. This was a very, very good start.

It wasn't all eureka moments in those first few months, and I have journal entries to prove it. In addition to the August 5 "Alex incredibly stupid today!" entry, I have many, many more: "Alex awful grumpy"; "One grouchy bird"; "Alex acts dumb today"; "Alex totally crazy this morning"; "Alex is totally impossible today, doing his war dance"; and so on. Maybe he had his reasons for these off days. I have no idea. But they became fewer as he became more confident, as we bonded as partners and built trust between us. We became less wary of each other. Nevertheless, for the first couple of years he remained extremely cautious with strangers, to say the least. He would shake, cower, and occasionally shriek. He often refused to cooperate with me when someone he didn't know was in the lab.

Yet he also began to assert himself with me. "Alex has become rather demanding if he's not promptly rewarded,"

I wrote on September 1. "After saying PAPER he repeats it more loudly and more quickly" if I was slow to produce it. It was as if he were saying, *C'mon, get moving, lady. I'm Alex. I want it now!* It was my first glimpse of a very different, more assertive personality that would soon emerge in full force.

When I arrived at Purdue at the very beginning of 1977, I knew exactly what I wanted to do. But I found myself in a quasi–Catch 22 situation. I needed grant money to support my research program, to pay for assistants, bird food, all the objects Alex was to label; to cover fees for my own laboratory space and maybe even a small salary for myself, but I didn't have a faculty position. It is very, very difficult—not impossible, but very difficult—to be awarded a research grant from the major funding agencies if you don't have a faculty position. At the same time, the authorities at Purdue said they might be able to give me a nonfaculty research position if I could get a grant. (It was made pretty clear to me that I was regarded as a faculty wife, David's, and that I should be content with that, rather than being a nuisance by trying to get a faculty position for myself, too.)

Nevertheless, I did secure a small piece of lab space in

which to do my work, kindly loaned to me by Peter Waser, an evolutionary biologist in the department of biological sciences. With a little artifice on my part with the dean, and the support of the department head, Struther Arnott, I managed to submit a grant proposal to the National Institute of Mental Health early in 1977, months before I even got Alex.

My proposal was simple: I said I wanted to replicate the linguistic and cognitive skills that had been previously achieved with chimps in a Grey parrot, an animal with a brain the size of a shelled walnut, but one that could talk. My confidence that I could do it was based on two things. The first was my experience growing up with talking birds, and the sense that they are indeed smart. Second were the facts that Greys, like apes, live a long time, and that their social groups are large and complex. Both these factors were thought to account for at least some of the brainpower that apes so obviously possess. Why not a similar kind of brainpower for Greys?

My plans for training Alex differed from the accepted standards of the time. Under the prevailing psychological dogma known as behaviorism, animals were seen as automatons, with little or no capacity for cognition, or thought. Biology was little better, dominated by theories claiming that much of animal behavior was innately programmed.

Experimental conditions for working with animals were very tightly prescribed. Animal subjects were actually supposed to be starved to 80 percent of their body weight so they would be eager for the food given for a "correct" response. They were also to be placed in a box so that the appropriate "stimuli" could be very tightly controlled and their responses precisely monitored. The technique was known as "operant conditioning." This was, to me, completely crazy, not to put too fine a point on it. It was contrary to all my gut instincts and commonsense understanding of nature.

For a start, isn't it blindingly obvious that communication is a social process, and that learning to communicate is a social process, too? It seemed clear to me that putting an animal in a box and expecting it to learn to communicate could not succeed. Several researchers had attempted to do this with mimetic birds and had failed spectacularly. They blamed the failure on a supposed deficiency in the brains of the birds, whereas I felt strongly that it was due to a deficiency in the researchers' assumptions and approach.

In fact, the first people working on human-animal communication in chimps in the late sixties and early seventies had not followed the behaviorism model. For the most part they had adopted much more naturalistic training techniques. Nevertheless, I still felt something was missing. And I couldn't quite treat a parrot the way the researchers were able to treat a baby chimp like a baby human, living with it 24/7 and still maintaining some objectivity. While mulling over this conundrum in 1975, I came across a paper by the German ethologist Dietmar Todt, published in what to me was an obscure German journal. In it he described his so-called model/rival program of training, which I adapted for working with Alex.

As I said earlier, under this system, instead of having one trainer, an animal subject had two. The principal trainer, A, would ask the secondary trainer, B, to name an object A showed to her. If B answered correctly, A would reward her; an incorrect answer would result in scolding. Trainer B is the "model" for the animal subject and its "rival" for the attention of trainer A. From time to time, trainer A would ask the animal subject to name the object, and it would be rewarded or scolded accordingly. Todt reported that Greys had learned speech very rapidly under this approach.

As soon as I read about Todt's work, I knew he was right, as far as he went. As promising as the approach was, I felt that one could not be certain that the birds understood the sounds they were using. To me, comprehension was key. If, for instance, Alex could produce a string of labels, no matter how clearly he enunciated them, it would be little more than mimicking if he didn't know that they were labels for specific objects or actions. I decided I would modify Todt's method, by, for instance, having trainers A and B alternate roles, so that the bird would learn that either role was possible. In addition, I would have the reward for a correct answer be possession of the object itself. If Alex were to correctly identify "paper," I or my partner would give it to him. Same with "key," "wood," anything. In this way the label and the object would become closely associated in his mind.

Bear with me as I wrap up this description of my training methods, using terms you'd be unlikely to hear in everyday descriptions of parrots learning words in people's homes. What I was planning to do wasn't an everyday exercise, of course. I was planning to demonstrate in a parrot cognitive processes that only humans and higher primates were considered capable of achieving. You need very special conditions to do that and, equally important, to have people believe what you might be claiming.

My training model would have three components. The first is *reference*, that is, what the word, or label, "means"; for example, the word "paper" refers to the physical object. The second, *functionality*, is the pragmatics of how the word is used; the reason to learn some odd set of sounds is that you can use it to get a specific, desired reward. The third is *social interaction*, that is, the back-and-forth, the relationship, between trainer and subject. The stronger the relationship, the more efficient the learning, just as with children. I always asked trainers to be enthusiastic in their exchanges with Alex and to emphasize the targeted labels, just as adults tend to talk to young children. With all this in place, we would have, I believed, the potential to explore the workings of a bird's brain as had never before been done.

Or at least that's what I argued in my grant proposal. Apparently, the review panel was not impressed. On August 19, just two weeks after the "by George, he's got it" moment, I received a letter from the panel that essentially asked me what I was smoking. They implied I was crazy to even imagine that a bird brain could master the language and cognitive skills I was hoping to demonstrate. And they further implied I was even crazier to shun the accepted approach of operant conditioning and adopt this highly suspect method of social interaction.

I shouldn't have been surprised. In retrospect, I was perhaps a little naive to expect the panel to give a grant to someone with no training and no qualification in psychology—or any biological science, for that matter— for a project that was at the very edge of what was then known and accepted. I was driven, however, and extremely confident that what I was proposing was going to work. So I *was* surprised, and very upset—so upset that Alex appeared to think from my behavior that I was angry with him. He cowered from me. "Oh, it's not you, Alex," I said to the poor guy. "It's those damned idiots who can't get out of their old ways of thinking. I guess we're just gonna have to try harder, buddy."

Nothing was going to stop me. Nothing was going to stop *us*. Alex and I pushed on with our work together, helped by Marion and a series of enthusiastic students. We introduced new objects—and new labels—to Alex, and he soon became a proficient, if occasionally recalcitrant, student. By the summer of 1978, a year after we started, Alex was demonstrating 80 percent accuracy in labeling seven objects and was beginning to learn colors, too, green and red (or rose, as we called it, to ease pronunciation). He was performing well enough on the strict tests we put him through that I felt I could reapply for a small grant, once again to the National Institute of Mental Health. All I was asking for was $5,000.

This time I was successful. In the pink summary statement I received that September, the panel described my proposal as "appealing." They said that "Alex is probably the best-treated parrot in captivity." Best of all, they concluded: "Approval was unanimously recommended." I was elated, of course, and did a dance of joy and relief. But there was a kicker: although the grant proposal was approved in theory, in practice there were insufficient funds available for me to receive any money. I was still in the same boat: no research funding, no research position. But at least I had Alex and his ever-growing list of accomplishments, and a few scientists had taken notice.

Onward we progressed, with more objects and another color, blue. I also introduced Alex to the concept of shape, which was related to number. A square, flat piece of wood we labeled "four-corner wood," and a triangle "three-corner wood." I struck a bargain with the guys in the woodshop at Purdue: they would supply me with four-corner and three-corner pieces of wood, and I would bake them cookies. In the absence of a grant to pay for such things, I had to be creative. In the end, the guys made the shapes from maple scraps, because Alex would destroy pine wood shapes in seconds; maple was much more of a challenge to chew to bits. And Alex loved challenges.

Along the way, Alex learned to say "no" and mean it. During our first year together, Alex had several ways of communicating displeasure or negativity of some kind.

Alex's First Labels

When he didn't want to be handled, for instance, he produced a high-decibel sound best conveyed as *raaakkkk*. He sometimes accompanied this extremely unpleasant noise with an attempt to bite, just in case his message had been misunderstood. When he didn't want to respond to a trainer asking him to identify an object, Alex would often simply ignore the trainer: he might turn his back or indulge in some suddenly urgent preening. He indicated that he had finished with his water or with a labeled object by simply tossing it on the floor. Give him banana when he'd asked for a grape, and you were likely to end up wearing the banana. Alex was not subtle.

Alex heard the word "no" a lot, from me or other trainers, when he incorrectly identified an object or was up to no good. By the middle of 1978 I noticed that Alex occasionally produced a "nuh" sound in situations where "no" would have been appropriate. "OK, Alex," I said, "why don't we train you to say it right?" Within a very few sessions, Alex replaced "nuh" with "no" in distress situations, such as not wanting to be handled. Very soon he used it to mean *No, I don't want to.* Here's an example of Alex with a well-developed sense of how to use "no." Kandis Morton, a secondary trainer, was working with Alex in April 1979:

K: Alex, what's this [holding a four-corner wood]?

A: No!

K: Yes, what is this?

A: Four-corner wood [indistinct].

K: Four, say better.

A: No.

K: Yes!

A: Three . . . paper.

K: Alex, "four," say "four."

A: No!

K: Come on!

A: No!

Alex was obviously in an especially obdurate mood that day, and was using "no" to express his unwillingness to go along with the training session. (He became even more creative in this respect as he grew older.) It was amusing, unless you happened to be the trainer trying to get some work done. Alex's use of the negative in this way represented a relatively advanced stage of linguistic development.

A few months after this session with Kandis, I had a set-to with Alex that provoked me to write in my journal: "Alex *definitely* understands NO!" By this time he had developed a passion for corks. On this particular August day he obviously wanted only the best of corks to chew. I gave him a new one. He contentedly proceeded to destroy it for a couple of minutes. When it was about two-thirds gone he dropped it. "Cork," he demanded.

"You have a cork, Alex," I said.

"No!" He picked up the sizeable remnant and tossed it on the floor. If he were human, I would have added that he did it with contempt. "Cork!"

I gave him a cork fragment, again sizeable but not complete. He snatched it from me, tossed it right back at me, and repeated even more urgently and impatiently, "Cork!" He would shut up only when I gave him a new, unblemished cork.

"This happened *all* morning," I wrote. I had wanted him to learn labels, and to express his wants. I guess I had succeeded.

Even at this early stage in our relationship, Alex was already showing that he was no birdbrain, no matter what the scientific establishment thought.

Chapter 4

||||||||||||||||||||||||||

Alex and Me,
the Vagabonds

One challenge I faced in trying to be taken seriously in this pioneering research was my complete absence of relevant publications. In academia, published papers are the measure of worth. I had several in chemistry, but those of course wouldn't count. By early 1979 I had enough good data on Alex's appropriate label use that I decided to submit a short paper to the American journal *Science*. The journal is very prestigious, so I was aiming high. But why not? Some of the very first papers on ape-human communication, by the Gardners, David Premack, and others, had been published there in the late sixties and early seventies. Why not the first one with a parrot?

I mailed the paper to *Science* in early May. It must have touched the editor's desk for a microsecond, because it came back immediately with only a short note saying it was not of significant interest. No comments. No helpful referee's suggestions. It obviously had not even been sent out to referees, just bounced right back. "I spend all day working on revisions, making calls, and being upset," I wrote in my journal on May 23. I also noted that a student, Gabrielle, was working with Alex on shapes: "Poor bird—he's really trying!"

If Alex wasn't giving up, then neither would I. The revision was for the journal *Nature*, the British counterpart to *Science*. The two journals are rivals, really, and don't always agree on issues and policy. But in my case, they were in lockstep: my submission came winging its way back, not refereed, just rejected again. I was crushed and felt awful. So, apparently, did Alex, though probably for different reasons. "Alex totally bitchy," I wrote in my journal. "Can't get any color marking at all—everything is ROSE; GREEN and BLUE don't exist. We can't even test him! Yuck!" It was just a bad day; he quickly got back into stride.

Alex could now identify objects we'd trained him to label, such as paper, wood, hide (rawhide), and key, and could label a limited set of colors. He was less interested in colors than the objects, probably because all the colors tasted the same, while the different objects had different tastes and textures. Now, could he correctly identify and name a novel combination of object and color—a blue key, for instance, whereas previously the colored keys he'd known were green, or blue objects that were something other than keys? In linguistics, this ability is known as segmentation, that is, being able to take pieces of two phrases apart and reassemble them appropriately.

I first tried this task with an old-fashioned wooden clothespin, what in England are called clothes-pegs. He loved chewing them. We called them "peg wood," which he picked up quickly. I then gave him a green clothespin, something he'd never seen before, and asked, "What's this?" He looked at it and cocked his head a couple of times, obviously intrigued, as he often was with novel objects. He then looked at me and said, "Green wood peg wood," all one phrase. We hadn't modeled it, so this was striking. Of course, a perfect response would have been "Green peg wood." But what he said suggested that he knew it all had to get together somehow, even if he wasn't quite sure how to do it. When we did model the correct response for him, he got it immediately. It was a wonderful start for something quite complex linguistically, in an animal with a brain the size of a shelled walnut. Very encouraging!

Even more encouraging was a letter on July 10. "I hear good news from NSF!" I wrote in my journal. "It looks like I will be funded for one year!" After my failed attempts with the National Institute of Mental Health, some colleagues told me that the National Science Foundation might be more interested in my research. I followed their advice, submitting a proposal early in 1979. And now I had succeeded. I was excited. I ran around shouting and clapping my hands. Poor Alex had no idea what was going on, of course, and was terrified by my wild behavior. "It's OK, Alex," I said. "Don't be frightened. We're funded. We're gonna be all right!" He did not look convinced.

My initial struggles with getting funded and published were happening at a time of growing controversy in the field of ape-human communication. Questions were being posed about its legitimacy. Leaders in the field— the Gardners, David Premack, Roger Fouts, Duane Rumbaugh and Sue Savage, Lyn Miles, and Penny Patterson—had employed a variety of methods for communicating with their simian subjects: hand signing in some cases, arbitrary symbols in others. The apes appeared to have demonstrated significant progress not only in labeling objects but also in creating novel phrases. The chimpanzee known as Washoe, for instance, in the care of Roger Fouts, had apparently coined the phrase "water bird" the first time she saw a swan; Koko the gorilla, the subject of Penny Patterson's research, seemingly described a zebra as a "white tiger." These efforts were garnering a lot of public attention (*NOVA* programs were just part of that; magazine and newspaper articles proliferated, too). Yet linguists were expressing a growing unease over the claim that these animals had demonstrated a rudimentary facility for language.

The subject of language has always been a contentious topic, scientifically but also emotionally. For both some scientists and laypersons, spoken language has long been held sacrosanct as being uniquely human, a defining character of what separates "us" (humans) from "them" (all other creatures). Too, a long-running debate exists about the more arcane issue of defining language. After all, other animals communicate with each other, often vocally. Is that not a form of language? I don't want to get bogged down in these controversies here. I just want to note that the rumblings of the gathering storm were getting louder and louder.

I was aware of the issues, of course, but not their scale when I initially set out on my journey. The first page of my Purdue journal naively proclaimed, "Project ALEX: Avian

Language Experiment." That's how Alex became Alex—not, as many people assumed, a play on "smart alec." His name was an acronym for where my research seemed to be heading: namely, I planned to develop parrot-human communication, using labels, as had been done with apes. That sounds a bit like language, doesn't it? And that was how the ape researchers had been expressing their goals and achievements. It was natural that I would follow their path.

But criticism began to mount, becoming ever more strident. Did the ape-language work have *anything at all* to do with language? The researchers might simply be deluding themselves—or worse, it was implied. I quickly realized that adopting terms the ape-language people were employing was probably unwise. It could distract from my real aim: namely, to explore the cognitive capacities of a nonhuman, nonprimate, nonmammalian animal, using communication as a window into his mind. I realized I needed to be careful about the terms I used in public and scholarly contexts.

About a year into the project, for instance, I started telling people who asked about the name Alex that it stood for "avian learning experiment," not "avian language experiment"—less provocative. In scholarly settings, I was ever more careful to describe Alex's vocal productions as "labels," not "words." And the post-*Science*, post-*Nature* draft of the paper would be titled "Functional Vocalizations by an African Grey Parrot." I thought it prudent. Words can be labels, and labels can be words. They can also be dangerous.

I sent the draft of my now lengthy article to a German journal, *Zeitschrift für Tierpsychologie*, in January 1980. A colleague had reminded me that it was where Todt had published his paper on the model/rival technique on which Alex's training was based.

Interestingly, about a month earlier, at the end of November 1979, *Science* had published a long paper by Herbert Terrace and several colleagues: "Can an Ape Create a Sentence?" It was to become a classic in the growing controversy. No one is more zealous than a convert, and Herb Terrace became such a convert. A leading psychologist at Columbia University in New York, Terrace had, until this point, been a strong advocate for the field of ape language, based on his initial research with an ape known as Nim Chimpsky (a play on the name of Noam Chomsky, the prominent linguist). Terrace's arguments were certain to be taken seriously. The *Science* paper was Terrace's mea culpa: his answer to the title's question was a resounding *no!* Terrace had analyzed Nim's hand-signing output in excruciating detail. He had expected to find evidence of grammar in Nim's supposedly spontaneous "utterances": instead, he said, the data were the product of unintentional cuing by his human handlers. That is, Nim was subtly following his handlers' leads, not spontaneously communicating on his own.

Terrace's *Science* paper was a huge blow to ape-language research, the first of two such blows in a six-month period. The second was even more devastating, both in its scale and in the sharp-edged language it employed to eviscerate the credibility of the field. It came in the form of a major conference, organized by linguist Thomas Sebeok and psychologist Robert Rosenthal, under the auspices of the New York Academy of Sciences in May 1980. The name of the conference was "The Clever Hans Phenomenon: Communication with Horses, Whales, Apes and People." It was a huge gathering of leading scientists, organized to denounce the work of the animal-language researchers: part of the bias that "they" cannot talk—only "we" can.

I eagerly attended the conference, partly because I hoped to meet some of the prominent researchers for the first time. It also introduced me to some of the younger people in the field, such as Diana Reiss, an expert on dolphin communication. Diana and I became instant and long-lasting friends. She and I were aware of the approaching assault on our field. But neither of us was prepared for the vitriolic atmosphere hovering over the audience in the elegant rooms of New York's Roosevelt Hotel during those extremely tense days.

Clever Hans was a German horse that performed in vaudeville acts in the 1900s. His owner, Wilhelm von Osten, would invite questions from the audience. Sometimes the questions were numerical, to which some number, say one to twelve, was the answer. Clever Hans would then tap out the answer with his hoof, stopping at the appropriate number. Clever Hans was a sensation. A horse that could add and subtract! A horse that understood questions in German! How did he do it? The answer, it turned out, was that when Hans reached the right answer, Osten involuntarily tilted his head a small fraction of an inch, and the horse detected it. The remarkable thing was that Osten was entirely unaware of his own head movement—it was an unconscious response. Without knowing it, Osten was cuing Clever Hans, whose cleverness was not arithmetic but highly developed visual perception.

It didn't take a detective to discern Sebeok's and Rosenthal's opinion of the field of animal-human communication, given their choice of conference title and what that implied. That the list of speakers included experts on training circus animals and a magician served to reinforce that conclusion. Sebeok and his wife, Jean Umiker-

Sebeok, circulated a manuscript prior to the conference in which they suggested that ape-language researchers had "involved themselves in rudimentary circus-like performances." A reporter for *Science*, Nicholas Wade, later wrote, "It's amazing that any of the ape-language researchers should even have considered stepping into such a lion's den." Actually, most did not. Only Duane Rumbaugh and Sue Savage showed up as speakers.

Diana and I sat openmouthed—figuratively, if not literally—as these academic icons flung themselves into battle. "Vituperative criticism" was how Sue Savage described the Sebeoks' assault, "replete with errors, both technical and logical." Their comments "embarrassingly reveal their incompetence," she added. Sebeok summed up his position in a postconference press gathering: "The alleged language experiments with apes divide into three groups: one, outright fraud; two, self-deception; three, those conducted by Herbert Terrace."

Whoa! I knew science could be contentious. But this? In Diana's report to her department she wrote: "One conclusion is to question whether scientists can communicate with one another, let alone with animals." In retrospect, I understood why editors at both *Science* and *Nature* had wanted absolutely nothing to do with my manuscript; they knew that this academic excoriation was coming

LEFT: Once more, for the camera: "What number is purple?" *Photo by William Munoz*

RIGHT: Now for some number comprehension: "What object two?" *Photo by Arlene Levin-Rowe*

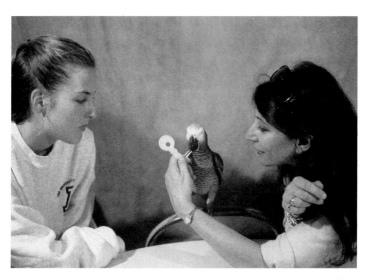

Alex in Arizona with Sandy Myskowski and Irene, demonstrating a model/rival session for a photoshoot. *Photo by William Munoz*

TOP: Alex, on left, helps train Griffin: "Talk clearly!" *Photo by William Munoz*

CENTER: An addition trial, but Alex is more interested in the photographer than the task at hand. *Photo by Arlene Levin-Rowe*

BELOW: Phoneme work, as in the (in)famous "n-u-t" incident: "What sound is blue?" *Photo by Jenny Pegg*

TOP: Alex on Irene's hand, eyeing his cage, as the camera flashes in his face. Immediately after the picture, he says, "Wanna go back!" *Photo by Karla Zimonja*

CENTER: Staging a number question for the camera. In a field of red, green, and blue wooden blocks: "How many red blocks?" *Photo by Arlene Levin-Rowe*

BELOW: Kyaaro (left), Alex (on hand), and Griffin (a baby at one year old) with Irene, mugging for the camera. *Photo by William Munoz*

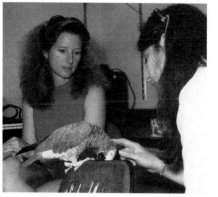

TOP: Kyaaro, with Diana May and Irene, asking for his favorite reward for a good session: "You tickle." *Photo by William Munoz*

CENTER: Alex on top of his cage with his excavated box, playing with sounds and practicing a new label. *Photo by Arlene Levin-Rowe*

BELOW: "Rose" is the answer to "What color smaller?" And Alex gets to chew it to bits. *Photo by William Munoz*

TOP: Yet again, for the camera: "What matter is green and four corner?" The answer is wood. *Photo by William Munoz*

CENTER: Alex in an iconic "mug" shot: The mug acted like an echo chamber, amplifying sounds when he put his head inside and talked. *Photo by Arlene Levin-Rowe*

Arthur (Wart) and Serial Tr-Hacking: Depending on which box has a nut, Arthur will have to "read" its arrow and pull the lever out or up, or give it a twist it to get the reward. *Photo by Ben Resner*

Alex has conned a student into tickling him instead of doing a session. *Photo by Karla Zimonja*

Alex checking out his daily breakfast of fresh organic vegetables and fruits. *Photo by Karla Zimonja*

TOP: Balancing all three birds for a "beauty shot"; Alex has to be closest to Irene's face. *Photo by Mike Lovett*

CENTER: Jesse (left) and Jessica praise Alex when he correctly answered the question, "What number green?" *Photo by Arlene Levin-Rowe*

BELOW: Alex's 30th hatchday, 2006: "Yummy bread!" *Photo by Arlene Levin-Rowe*

Alex, on "the chair," keeps Irene company as she checks e-mail. *Photo by William Munoz*

Alex on his training perch, resting between trials. *Photo by Karla Zimonja*

over the horizon. *Thank goodness I got my NSF grant before all this hit*, I thought. *Thank goodness my new manuscript is in the hands of editors of a journal that see merit in asking questions about animal thinking and can see value in our methods.*

I returned to Purdue after the conference, still stunned. I went into the lab and heard Alex's now familiar "Come here" greeting as I approached the curtain that partitioned off his area. I pushed it aside, and there he was, waiting for me. Then he added something he had begun to say from time to time: "I love you," learned from the students. I went to his cage, where he was perched on top, displaying excitement at my return. He raised his wings slightly and lifted a foot, so I offered my hand as a perch. "Thank you, Alex," I said, as he climbed onto my hand. "What have we got ourselves into, buddy?" He didn't seem concerned. He happily preened himself.

We had developed a degree of intimacy by this time, of course. We often spent eight hours a day together. But from the very start of The Alex Project I had determined that my professional approach would be rigorous in training and in testing my Grey. I had come from the so-called hard sciences, after all. I needed my data to be unimpeachable, to meet high standards of credibility. I wouldn't let emotion cloud my judgment. I wouldn't get *too* attached. My experience at the Clever Hans battle made me even more determined to maintain as much of an emotional barrier as was feasible between Alex and me in order to keep that credibility intact, no matter how hard it would be. And it was hard.

Alex and I were like vagabonds during our seven-plus years at Purdue, lugging our limited belongings from temporary lab to temporary lab, constantly in search of more permanent space. We never found it. There was an amusing biblical aspect to our wanderings, too. We coped with more than one laboratory flood, which required evacuating a freaked-out Alex from rising waters in the middle of the night. And pestilence: the cockroaches were always awful. No matter where we were, all the rooms adjacent to ours were regularly sprayed for roaches, but we couldn't do that, for fear of poisoning Alex. As a result, we seemed to be a haven for bands of refugee roaches. We had to literally vacuum them out of drawers every week and spray the floors with alcohol. We put stickies around Alex's cage in an attempt to trap the beasties. It didn't always work very well, and sometimes there were roaches in his water in the morning. Alex hated it as much as we did.

When news of the NSF grant had arrived back in 1979, it had provided us with a little stability, inasmuch as I finally got a real position: a lowly research associateship for one year. But things were beginning to look up. I began to give presentations about our work at local and national animal behavior meetings. The NSF grant was extended for a second year; my big paper was published in the German journal early in 1981. Reaction among my peers was, I have to say, rather muted. But it did prompt the beginning of public recognition: first in *Omni* magazine, then a snippet in the *New York Times*, followed by a piece in the popular journal *Science 82*; our local TV station did a spot on Alex, too. More people in the department were willing to openly support our "edgy" work and ideas, but there were always more or less vocal detractors.

There would be bumps ahead, but we were moving in the right direction. Together, Alex and I would show them what it really means to be a birdbrain.

We had already shown that Alex could appropriately label objects, which he wasn't supposed to be able to do; that he could correctly label colors, which he was not supposed to be able to do; and he had a functional use of "no," which he was not supposed to be able to do.

He was also on his way to understanding concepts—such as color and shape—at a higher level of cognition. It was one thing for us to ask Alex to identify an object and for him to correctly say "green key" or "four-corner wood." But it was quite another for him to see a piece of blue three-corner paper or red four-corner hide and correctly respond to separate questions such as "What color?" or "What shape?" (Actually, we had to give up on the paper objects, because when he chewed them the vegetable dyes came off on his beak and transferred to his feathers, his feet, his perches, and ultimately his trainers—a colorful mess.)

To answer "What color?" and "What shape?" correctly, Alex needed to understand the concept of color and shape as *categories* that *contained* the labels "green," "blue," "three-corner," and "four-corner," not simply labels by themselves. He passed that test during his third year of study, and, of course, he was not supposed to be able to do that, either. This was grist for another scientific paper, published in 1983. So far, everything that Alex was not supposed to be able to do, because he was a mere birdbrain, he had done.

Meanwhile, Alex was up to all kinds of mischievousness in the lab, which he was not supposed to do, either. Trouble is, Greys just love to chew things. It's their nature. But Alex, being Alex, loved to chew *important* things, such as telephone cables (thus disabling two professors' lines as well as mine) and my lecture slides, in the days when it took weeks of labor to prepare them. And more. Back in

1979, he had a go at the grant proposal to NSF that got me my first funding. The panel had obviously been impressed with it, but that was only after Alex had expressed his own avian opinion. I had spent the previous night and the whole of the morning putting the finishing touches to this twenty-page document that carried so much of my hopes, crafted on a borrowed electric typewriter. I stacked it neatly, put it on my desk, and met a colleague for lunch. Big mistake.

I returned to find most edges heavily chewed. It was unsalvageable—I'd have to retype the whole thing. Damn it! I had only a few hours to photocopy it and mail it out. I responded irrationally, as humans often do in such circumstances: I shrieked at Alex, yelling stupidly, "How could you do such a thing, Alex?" Easy—he's a parrot.

Alex then employed something he'd learned recently in similar circumstances. He cowered a little, looked at me, and said, "I'm sorry . . . I'm sorry."

That stopped my ranting. I went to him and apologized. "It's OK, Alex. It's not your fault."

How had Alex's use of "I'm sorry" come about? Shortly before the grant-chewing incident, Alex was on a high perch and we were just hanging out, chatting, nothing formal. I was drinking coffee. He was preening and making contented noises. I put my cup on the base of the perch and went to the washroom. I came back to find Alex wading in spilled coffee on the floor among shards of broken coffee cup. I was panicked, afraid he'd hurt himself, and yelled, "How could you do that?" Alex must have simply knocked the cup off the base as he went to investigate—an accident. But I yelled anyway, until I realized that I was the stupid one, doing the yelling. I got down to make sure he was OK, and said, "I'm sorry . . . I'm sorry." He obviously learned that "I'm sorry" is associated with defusing a tense, angry, and potentially dangerous moment. That's why he applied it to the grant-chewing incident, when I was again stupidly yelling at him. Who is the birdbrain?

Alex became more subtle in his use of "I'm sorry." Alex was wonderful with training and testing when he wanted to be, and not when he wasn't. Usually when he was uninterested in working, he would ignore us, preen, or say "Wanna go back," meaning he wanted to go back to his cage. Late in March 1980, however, he did something new. I and Susan Reed, a student, were trying to test him. Alex was completely recalcitrant, refusing to do anything. "Alex wouldn't test," I wrote in my journal. I was a little pissed off, probably in a bad mood myself, and abruptly started to leave the room, annoyance evident in every aspect of my body language. Next thing I heard was "I'm sorry."

It was Alex. I went back in. *Hmm*, I thought, *does he mean that?*

A little later that morning, another student, Bruce Rosen, was working with Alex, playing with a plastic cup. Alex accidentally knocked it to the floor. Alex wasn't aware that I was watching him. And he said again, to Bruce this time, "I'm sorry." I went to him and said, "It's OK, Alex; it's all forgiven."

I wrote in my journal that evening, "Does he understand?" I meant, did he feel remorse, such as you or I might feel when we say "I'm sorry"? Or was it simply a means of defusing anger? Either way, it was an effective mode of communication. As he grew older, he began to say "I'm sorry" in an ever more pathetic, "I'm really, really sorry" tone of voice, which always had the effect of melting my heart, no matter how he meant it.

Ever since the phone wire chewing incident, students were instructed not to leave Alex alone in the lab. He simply couldn't be trusted not to get into trouble, no matter how short the window of opportunity. Sometimes they would put him in his cage if they had to leave for a while. He didn't much like that. When it was a quick trip to the washroom, and when he had pretty much lost his fear of strangers, the students would sometimes take him with them. He definitely liked that, especially if someone else came in so he could show off, whistling or talking, "Want nut," "Want corn," and so on.

Now, these trips to the washroom brought up another issue—but first I must digress. Fairly early on I had planned to use a two-way mirror in the lab for observing Alex unseen by him. Alex's cage was supposedly angled so that he would not see himself. But such was not to be. "Introduced Alex to the 'bird in the mirror' today," I wrote in my journal. "What a flaky parrot—he's truly scared of himself." We obviously can't know what he was thinking. But when I pulled back the screen that had until this time covered the mirror, all of a sudden there appeared to be a window in the room. Alex looked, saw "another bird," and was visibly scared. "He actually crawled to me for comfort," I wrote, "which shows how freaked out he was." I doubt that, from his viewing angle, he could really have made any connection between himself and this other creature; even to me, it looked like another room with another bird.

As time passed Alex grew less timid with the situation. That was a good thing, because the washroom where the students occasionally took him had a very large mirror above the sinks. Alex used to march up and down the little shelf in front of this mirror, making noise, looking around, demanding things. Then one day in December 1980 when Kathy Davidson took him to the washroom, Alex seemed really to notice the mirror for the first time. He turned to look right into it, cocked his head back and forth a few times to get a fuller look, and said, "What's that?"

"That's you," Kathy answered. "You're a parrot."

Alex looked some more and then said, "What color?"

Kathy said, "*Gray*. You're a *gray* parrot, Alex." The two of them went through that sequence a couple more times. And that's how Alex learned the color gray.

We don't know what else Alex learned from the mirror that day, what thoughts were in his mind as he saw his reflection in the mirror. But it did mean that formal mirror tests were now impossible.

Chapter 5

|||||||||||||||||||||||

What's a Banerry?

The July 4 holiday in 1984 was our last day at Purdue, our last day in West Lafayette, Indiana. Movers packed our belongings and I packed the lab, Alex included, into a rented station wagon. A student and I drove the 120 miles to the town of Wilmette, Illinois, about sixteen miles north of downtown Chicago, on Lake Michigan. David's post at Purdue had ended, and we'd had to move on. He obtained a faculty position at the University of Illinois–Chicago, and I had snagged a temporary, one-year appointment at nearby Northwestern University, in Evanston. We drove at night, hoping that Alex would sleep through the trip. It was his first time in a car since the original stressful drive from Chicago. But

Alex was awake and alert the whole time. He held on to the side of his cage with one claw, looking just like a New York City strap-hanging subway rider.

We had left behind the endless miles of cornfields and successions of summer tornadoes that, for me, defined West Lafayette. Alex had been terrified by the tornadoes. He could sense the change in air pressure long before we were aware of anything: the only thing that soothed him as the storms raged was Haydn's cello concerto, which sometime swept him into a trancelike state, his body moving gently, eyes squinting almost shut.

Before us lay a new life in Illinois, and a new Alex. Alex was no longer anxious and skittish around strangers. Quite the opposite. He had learned that by asking for things he was in control of his environment, because when he asked for things we gave them to him. He liked that. He let it be known to newcomers very clearly and very quickly that they had to obey, as my friend Barbara Katz discovered on her first encounter with Alex. Barbara, who was in charge of birds at Lincoln Park Zoo, and I met in a doctor's office in Evanston soon after I arrived in town. We became fast friends.

Very shortly afterward I had to leave town for a conference in Boston. I asked Barbara if she would look in on Alex and the students while I was gone. She said she would be delighted. I was confident that her years of work at the zoo would equip her to handle Alex. Here's how she later described the encounter:

I was an experienced bird handler and I thought it would be a simple matter.

I arrive at the lab early in the afternoon to find Alex happily destroying an old wooden cabinet while the students sat helplessly in chairs a few feet away.

"Hi Alex. How's it going?"

"I want walnut," he declared in his charming sing-song voice.

"Alex," I said gently, "you eat too many nuts. Irene said to offer you fruit if you started asking for treats. How about a grape?"

"I want walnut."

"No walnuts. How about some banana?"

"I want walnut!"

"Okay, just one."

I removed one walnut piece from a metal tin and held it in my outstretched hand. Alex deftly reached over and took it. He nibbled it, eating tiny pieces until it was all gone except for some telltale crumbs at the sides of his beak.

"I want walnut."

"No, you just had one. How about a grape?" I sensed trouble ahead.

"I want water."

"That's a good idea, Alex." I held out his small white plastic cup. Alex drank two sips, then pulled the cup from my hand and contemptuously tossed it on the floor.

This was the new Alex. The boss.

Aside from being "pretty crazy for several days" immediately after our move, Alex was becoming more and more confident. Within less than two weeks, he was responding correctly to "What color?" when I showed him a three-corner gray wood. "Gray," he said. Then "Gray wood." "This after a long break, IN NEW LAB, NEW PEOPLE and NO TRAINING," I wrote in my journal. The not-so-subtle emphasis in my writing expressed my delight at Alex's supremely capable performance.

When David learned in the summer of 1983 that he would be moving to the University of Illinois, I'd had to scramble for a job. At one point I faced the prospect of taking Alex to the University of Massachusetts in Amherst, where a friend offered some lab space for Alex and me for a year. Not ideal. At the very last minute, a one-year slot opened up in the anthropology department at Northwestern University for a visiting assistant professor who could teach animal behavior. I grabbed it. I remember thinking, *Hey, you have a job. OK, it's only a year and there's no chance for tenure, but it's a job. You have some grant money. You are being paid to teach. Not bad!*

The anthropology department's labs were in Swift Hall, on the north end of campus, on the lake. Physically, Northwestern's campus was gorgeous. I was given one room on an upper floor of Swift Hall for Alex and a little office on the ground floor. There wasn't much in the lab, just the ratty desk, on top of which sat Alex's cage, and a small folding metal chair on which Alex loved to perch. My office had a desk, a bookcase, and a chair. The ceilings were high. Overall, the place felt rather dungeonlike, as a friend put it. But it was *our* space, and Alex and I made good use of it.

A few months after we moved in, a student volunteered to help in the lab. In exchange, I offered to train his parrot, who at that point was wordless. The parrot's favorite thing in the world was apple, so we decided to train her to produce the label "apple." Alex would take part, too. We had never before used food items as training objects with Alex, so this was going to be an exception. Alex had acquired "grape," "banana," and "cherry" on his own, because we named everything we fed him. "Apple" was therefore going to be his fourth fruit label. Or so we thought. Alex apparently had other ideas.

By the end of the season for fresh apples, Alex had learned to produce a puny little "puh" sound, a pathetic fragment of "apple." Nothing more. And he entirely refused to eat apple. We decided to try again the next spring, when fresh apples would arrive from the Southern Hemisphere. Months later Alex did condescend unenthusiastically to eat some apple when offered, but still only produced "puh."

Then suddenly, in the second week of training in mid-March 1985, he looked at the apple quite intently, looked at me, and said "Banerry . . . I want banerry." He snatched a bite of the apple and ate it happily. He looked as if he had suddenly achieved something he had been searching for.

I had no idea what he was talking about. So I said, "No, Alex, *apple.*"

"Banerry," Alex replied, quickly but quite patiently.

"Apple," I said again.

"Banerry," Alex said again.

OK, buddy, I thought. *I'll make it a bit easier for you.* "Ap-*ple,*" I said, emphasizing the second syllable.

Alex paused a second or two, looked at me more intently, and said, "Ban-*erry,*" exactly mimicking my cadence.

We went through this double act several times: "Ap-*ple.*" "Ban-*erry.*" "Ap-*ple.*" "Ban-*erry.*" I was a little ticked off. I thought Alex was being deliberately obtuse. In retrospect, it was quite hysterical. When I told one of my students, Jennifer Newton, about it later, she literally fell off her chair laughing. But Alex hadn't quite finished with me just yet. At the end of that session he said, very slowly and deliberately, "Ban-*err*-eeee," just as I might do with him when I was trying to teaching him a new label. Maybe he was thinking, *Listen carefully, lady. I'm trying to make this easy for you.* I wrote in my journal that Alex seemed "almost angry with us."

I still had no idea what Alex was talking about, even though he obviously thought he did. Try as we might, he wouldn't budge from "banerry." No matter how hard we worked to get him to say "apple," he stuck with his label. As far as Alex was concerned, "banerry" it was and "banerry" it was going to stay.

A few days later I was talking to a linguist friend about all this. He said, "It sounds like lexical elision." It's a fancy term for putting parts of two different words together to form a new word. Alex might have thought the apple tasted a bit like a banana. Certainly it looked like a very large cherry (it was a red apple). "Banana" + "cherry" = "banerry."

Had Alex done this intentionally? It certainly seemed so, but intentionality is a hot-button issue in animal behavior circles, and proving it is very difficult. Alex often played with sounds, particularly when learning new labels, and especially when he was by himself in the evening. In contrast, these novel sounds typically were nonsensical. And until this point Alex had not said "ba-nerry" in any session with an apple, nor in any informal setting. It really did look like some bird brain creativity of a sort never previously seen. Of course I can't document that scientifically. I can't report that he'd actually gone and decided that that's what he was going to call an apple, and that he was not going to change his mind. It had to remain something remarkable just between Alex and me.

The original proposal for The Alex Project that I wrote back in the spring of 1977 had been, I must concede, quite ambitious. It argued that my Grey would learn object labels (words), categories, concepts, and numbers in three years; that he would be able to communicate back and forth with a human; and that he would have some comprehension of what he was doing. I had complete confidence that my Grey would be able to do all this. But I have to admit that each time Alex met a challenge I set him, each time he did what no bird brain is supposed to be able to do; I was as thrilled as any parent can be when her child crosses a developmental hurdle, such as crawling, walking, or talking.

As the list of scientific publications grew—and as our work garnered more and more public attention—I found a slowly growing acceptance that I wasn't just "that woman who talks to a parrot." I was beginning to be taken seriously in scientific circles. But the chorus of "Oh, he's just mimicking" or "He's just following her cues" still sounded loudly in my ears. At least that is how I perceived it. I found myself having to prove over and over that Alex had more going on in his bird brain than some mechanical trickery or other. One such challenge was, "Oh, he can produce labels all right, and he *sounds* convincing, but does he really understand what he's saying? Does he comprehend the noises coming from his beak?"

It seemed quite clear to me from my hundreds and hundreds of hours watching and listening to Alex that he did indeed know what he was saying. A simple example: if Alex said "Want grape" and you gave him banana, he'd spit it right back at you and repeat insistently, "Want grape." He wouldn't stop until you gave him a grape. If you were dealing with a child, you would accept without question that he or she really wanted a grape, and that banana simply wouldn't do. But that's not science. Science needs numbers. Science needs tests to be done over and over again—actually, sometimes sixty times or more—before the answer has statistical legitimacy, and before scientists will take you seriously. Poor Alex.

A few years into our Northwestern period—my initial, temporary job ultimately stretched out to six and a half years—we embarked on a rigorous series of tests of Alex's comprehension ability. Scientifically I can report that he passed each test, and move on to the next part of our story. But *how* he did it gives us insights into his mind that are striking, if not always quite so easily classified as scientific.

The tests involved putting various of his "toys" on a tray and asking questions such as "What object is green?" "What matter is blue and three-corner?" "What shape is purple?" "How many four-corner wood?" At first, Alex answered correctly most of the time: "key" or "wood" or "wool" or "three," et cetera. But before too long, he started to act up. He would say "green" and then pull at the green felt lining of the tray, hard enough that all the objects would fall off. Or he would say "tray" and bite the tray. Sometimes he'd say nothing and suddenly start preening. Or he'd turn around and lift his butt in my direction, a gesture too obvious to need translation. Once he grabbed the tray out of my hand and flung it on the floor, saying, "Wanna go back," which meant, *I'm done with this. Take me back to my cage.*

Who can blame him? None of the objects were new to him. He'd answered these kinds of questions dozens of times, and yet we still kept asking them, because we needed our statistical sample. You could imagine him thinking, *I've already told you that*, *stupid*, or simply, *This is getting very boring*. He was like the bright little kid at school who finds none of the work challenging and so passes the time by making trouble.

Sometimes, however, Alex chose to show his opinion of the boring task at hand by playing with our heads. For instance, we would ask him, "What color key?" and he would give every color in his repertoire, skipping only the correct color. Eventually, he became quite ingenious with this game, having more fun getting us agitated rather than giving us the answers we wanted and he surely knew. We were pretty certain he wasn't making mistakes, because it was statistically near to impossible that he could list all but the correct answer. These observations are not science, but they tell you a lot about what was going on in his head; they tell you a lot about how sophisticated his cognitive processes really were. Whether you would describe what he did as something to amuse himself or as making a joke at our expense, I cannot say. But he was definitely doing something other than routinely answering questions.

We became ever more ingenious in presenting our questions, to keep a step ahead of his boredom. Sometimes we succeeded, sometimes we didn't. In the end we did arrive at a statistically valid answer to the question "Does Alex know what he's saying?" Yes, he did. His level of comprehension was equal to that of chimpanzees and dolphins, no small achievement for so small a brain.

Alex faced the same boredom with the next major challenge we gave him: namely, would he understand the concept of "same" and "different"? It might seem like common sense that in order to survive in the wild, birds would, for example, have to identify the songs of individuals and distinguish among species. Surely this involves some grasp of "same" and "different." Yet when I embarked on the "same/different" project with Alex, the scientists who test such things thought that apes were at par or slightly below humans in this conceptual ability, monkeys were below apes, and birds . . . well, they hardly counted at all.

The concept of "same/different" is fairly sophisticated cognitively. We trained Alex to use color and shape as categories with which to determine same or different. When presented with a pair of objects, such as green four-corner wood and blue four-corner wood, Alex's correct response to "What's same?" and "What's different?" would be "shape" and "color," respectively, not the specific color or shape. To answer the question correctly, Alex would have to take note of the various attributes of the two objects, understand exactly what I was asking him to compare, make that judgment, and then vocally tell me the answer. No small order for a bird brain.

It took many months to train him, but eventually he was ready to be tested. Because many of the objects we used were familiar, boredom again became an issue. We tried to keep his interest by interspersing "same/different" tests with teaching him new numbers, new labels, and other novel tasks. He was a trouper. Overall, he got the right answer—"shape" or "color"—about three-quarters of the time. (We also included a third category, "matter," or material.) When we gave him pairs of objects that were novel to him, colors he could not label, for instance, he was right 85 percent of the time, which is actually a better measure of his ability. The novelty obviously held his attention better.

Now, when David Premack had tested chimps on this kind of test, all the animal had to do was indicate whether two objects were the same or different. Alex went a step further in our tests. He was able to tell me exactly *what* was the same or different: color, shape, or material. When I reported our results at the International Primatological

Congress in Göttingen, Germany in 1986, a senior prima-tology professor—we called them "silverbacks," referring to the markings of older gorillas—lumbered to his feet and said, "You mean to tell me that your parrot can do what Premack's chimps can do, only in a more sophisti-cated manner?"

I said, "Yes, that's right," wondering what onslaught might follow. Nothing. He simply said, "Oh," and sat down. I might have burst into song, with "Anything chimps can do, Alex can do better," but I restrained myself. Besides, my voice isn't up to it. Nevertheless, this was a moment of triumph for Alex. Pity he wasn't there to witness it.

From the "same/different" challenge, it was natural to go on to relative concepts, such as size difference. Alex got that, too. I could show him two different-sized keys, each a different color, for instance, and ask him, "Alex, what color bigger?" and he could tell me. These vari-ous achievements attracted a lot of public attention. Bob Bazell, of NBC television, came to film Alex, as did crews from ABC and CBS. Alex was even on the front page of the *Wall Street Journal. Very* smart bird!

My time at Northwestern had begun wonderfully: a job, grant money, amazing results with Alex. It didn't last. In the summer of 1986, I learned that my NSF grant request had been approved but, as had happened before with NIMH, there were no funds to give to me. I feared I might have to leave Northwestern. The department head told me he would have to find someone else to teach the animal behavior course: without the overhead from my grant, there was no money to pay me. My marriage, which was already rocky, came under even more strain. David essentially told me, "You're a failure. Why don't you close the lab and go get a real job? We need the money to live here in Chicago."

I was angry, a barely contained volcano. Nothing could have focused me more than to be told I was a failure and that I should give up what had become my life's work, give up Alex. I frantically searched for somewhere to continue, contacting friends, acquaintances, and colleagues all over the country. Friends in Kentucky told me they might have an opening for me there, but only for a year. I lost thirty pounds in three months. My friends were my only comfort, aside from Alex.

I was now spending virtually all my time in the lab, including evenings. Alex and I hung out together at the end of each day, me trying to make plans, Alex preening, the two of us occasionally exchanging words—not exactly a conversation, but as close as it gets with a nonhuman companion. Like all Greys, Alex was very empathic. He could sense when I was particularly blue. He would sit close with me at these times, just being Alex. Not Alex the mischievous imp; not Alex the boss of the lab; not the demanding Alex. Just Alex the empathic presence. He'd sometimes say "You tickle," and bend his head so I could scratch his face. As I did, the white area around his eyes turned a subtle pink, blushing as Greys do when being intimate. His eyes would squint almost closed.

Just as things seemed darkest, with a week to go before classes began, I was told that the department couldn't find anyone else to teach animal behavior after all, and the job was mine if I wanted it. *If I wanted it?* Reprieved, but still without grant money, I ran the lab on a shoestring for a year. The students were now volunteering because I could no longer pay them. I had resubmitted my grant request; it was again approved, and this time it was fully funded. It had been a very hard twelve months.

That wrenching episode proved to be the gateway to three years of extremely productive work for me at Northwestern. I worked with Alex on numbers, on why our particular training technique was so effective, and on how it related to what birds do in the wild. I collaborated with Linda Schinke-Llano, a specialist on second-language acquisition in humans, looking at how Alex's learning of

English words illuminated that process. We teamed up with people in biological sciences, showing how birds' acquisition of other birds' songs also resembled second-language acquisition. My students and I did some preliminary work on so-called object permanence: that is, the understanding that an object continues to exist even when it is hidden. This capacity develops gradually in children over their first year. Alex clearly had a good grasp of it and used it to have fun with our testing procedures.

Denise Neapolitan, a student, and I did a little study that asked whether people talk differently to male and female parrots. Alex played both himself, the male, and "Alice," the female, in this test. The answer was almost predictable: people used more "baby talk" when talking with "Alice" than with Alex. Another student, Katherine Dunsmore, received a small grant with which to buy recording equipment. We taped Alex's evening babbling, when he was free to "practice" sounds and new labels before going to sleep, just as children do. Ruth Weir had published a now classic book, *Crib Talk*, on babies back in 1962. When Kathy and I wrote our paper, we wanted to call it "Cage Talk," but the editors wouldn't let us. Instead, we had to call it "Solitary Sound Play During Acquisition of English Vocalizations by an African Grey Parrot." Boring, but accurate.

All in all, I had a lot going on in my professional life at that point, partly because that is the kind of person I am, but also, I suspect, as a substitute for what was missing in the rest of my life. I was living a divided existence: much of me was relishing the exhilaration of the great strides I saw Alex making in our work together; the rest of me was an empty ache.

Since my arrival at Northwestern, I had been applying for regular university positions, but after the scare in 1986, I began to do so in earnest. Not much had come my way, except what I called a few "affirmative action" interviews—I was clearly the lone woman called, and during the interview, one or more comments would make it clear that I was not being considered seriously. I wasn't too concerned, at least until the fall of 1989, when Northwestern informed me that my position as a visiting assistant professor could be renewed at most until the end of 1990—not because Alex and I hadn't done well, but because there were rules about temporary positions. David's response was similar to what he'd said in 1986—why couldn't I get a "real" job? Another round of applications, another round of interviews, and plenty of stress. This time I was offered a job that could lead to tenure at the University of Arizona, in Tucson, in May 1990. For a number of reasons, I decided to put off the move until the last possible moment, at Thanksgiving.

Meanwhile, Alex continued to attract attention from local and national television. He seemed to enjoy performing for the media and was not at all camera-shy. We had many other visitors, too, one of whom stands out because the event was so nerve-wracking, at least for me. One day in the early fall of 1988, my friend Jeanne Ravid asked if she could bring someone to the lab to meet Alex. Jeanne explained that her friend was in town for a short while, staying at her house in Evanston, just south of campus. "Garrick loves Greys," she said, "but he can't have one as a pet because he travels so much. Garrick always stays with me when he's in Chicago," she continued, "partly because we have a grand piano, so he can play when he wants to, but also because of our Grey, Wok. He loves Wok."

I knew Wok because he had taken part in Alex's object permanence study. And I knew Jeanne's house, too: big, with a beautiful grand piano in the elegant front room. Something rang a bell: *Hmm, Garrick, plays the piano, travels a lot . . .* Jeanne explained that Garrick had heard of my work with Alex and knew that Jeanne and I were friends.

"Wait a minute, Jeanne," I said, "do you mean the pianist, *the* Garrick . . . ?"

"Yes," she said, "Garrick Ohlsson." Ohlsson was the first American to win the International Chopin Piano Competition, in 1970, and had a towering reputation in the world of classical music. It would be a privilege to meet him, but I was thinking, *Oh, my God! I can see the headlines now: "Internationally Renowned Pianist Loses Finger to Parrot." Please, Alex, don't do anything awful.*

Jeanne brought Garrick into the lab the following day. He is a big man, standing more than six feet, with a trim, square beard and a great presence, a true star. But with Alex, he was like a kid on Christmas morning, such was his excitement at meeting my own star. Alex behaved wonderfully. He happened to like guys a lot, particularly tall ones, and he seemed thrilled to meet Garrick. He jumped on Garrick's arm, scampered up on to his shoulder, and did his "I'm really happy to be with you" dance, which is actually part of a Grey's courting ritual. Garrick was thrilled, too. He left with ten fingers intact. And we had tickets for symphony hall that night.

Not too long before we left Chicago, Alex gave me a terrible scare. In early September 1990, I returned from a short trip and found a message from a student on my answering machine. "I had to take Alex to the vet today, because he was wheezing badly," it said. "Call the vet immediately." I did, right away.

"Susan, what's wrong?" Susan Brown was one of three partners in the vet practice in a western suburb of Chicago that I routinely used.

"It's another one of these cases," she said. "Aspergillosis." Aspergillosis is a fungal infection that can affect the chest cavity and the lungs. Alex probably caught it from contaminated corncob bedding placed at the bottom of his cage in my absence when the usual pine shavings weren't available. Several cases had turned up at local vet practices in the previous few weeks. "Look, it's not bad," Susan said, trying to calm me. "He'll survive. I'm at the movies. I'll call you when I get out." Susan had one of the first cell phones. It looked like a brick and weighed as much as one.

I went straight to my bird medical book, looked up aspergillosis, and froze. "Make your bird comfortable and wait for death" was the essential message. I was now panicked, and could barely contain myself until Susan called. She again reassured me, saying that the book was out of date and that Alex was going to be OK. "Trust me," she said. "I'll give you some meds for him. Come get them tomorrow."

I medicated Alex for about a week in the lab, but he was getting no better. I talked with Susan every day. Eventually she told me to bring him in so they could try some new drugs. Apparently, the outlook wasn't as rosy as Susan had implied, because treatment for aspergillus infection in Greys at that time was not well developed. The one vet who specialized in the disease had developed drug regimes for raptors: eagles and other birds that weighed twelve times as much as Alex. Susan told me that she and her colleagues would have to experiment with the drug dosage, so Alex would have to stay in the hospital for a while.

When I was getting ready to leave, I said, "Goodbye, Alex." He looked at me, obviously feeling lousy and frightened, sitting in a small veterinary cage. "I'm sorry," he said in a small voice. "Come here. Wanna go back." He sounded so pathetic, it really tore me up. "It's OK, Alex," I said as reassuringly as possible. "I'll see you tomorrow. I'll be back tomorrow." I had been saying that to him for a while, but this time it really meant something. It was important for him to understand that I was going to come back and that I wasn't just going to leave him there.

Each day I got up early, went to Northwestern, taught my class, and then drove an hour to the vets' office, to spend as much time as I could with Alex. I'd leave by three o'clock to avoid the rush-hour traffic, return to the office and lab, pack for the upcoming move to Tucson, and eat a sandwich or such for dinner. It was very wearing, physically and emotionally.

The vets administered some of the drugs by putting Alex in a nebulizer, essentially a tank into which they would vaporize the drugs, so he would breathe them in. The poor bird hated the process and at first kept saying, "Wanna go back . . . wanna go back." Each time he had to wait until a timer went off, signaling the end of the session. He soon learned the routine, waiting until the bell rang before demanding, "Come here. Wanna go back!"

On one occasion there was an emergency at the office, and the vets couldn't get to Alex immediately after the timer went off. "Wait a minute, we're busy," they called to him. Alex couldn't wait. After a couple of futile "Wanna go back" demands, he started to rap his beak on the glass wall of the nebulizer. "Pay attention!" he said. "Pay attention. Come here. Wanna go back!" He had picked up the phrase from students who, during training, would say to him, "Come on, Alex, pay attention!"

By early November, it was obvious that the meds weren't working. Susan and her colleagues, Richard Nye and Scott McDonald, called me in for a conference on how to proceed. One option was to continue with the drug regime and hope it would eventually work. Another was surgery. But that was experimental and could be dangerous. Finally I said I wanted to try surgery.

At that time, there were only two veterinarians in the country who could do the microsurgery we needed. It involved scraping fungal spores out of Alex's chest cavity. One was Greg Harrison, in Lake Worth, Florida, whom I had met at a conference. Fortunately, I had gone through all the red tape necessary to have a bird fly with me in a plane cabin, for the trip to Tucson. All I had to do now was get the tickets. I cancelled a talk I had scheduled with the Northern Illinois Parrot Society; unbeknownst to me at the time, they held a fund-raiser instead, to help out with the cost of Alex's treatment. I put my own ticket on a Visa card. I called Ernie Colazzi, a friend who had once said to me, "If there's anything you need, anytime, just call," and explained I needed $600 for a plane ticket for Alex. He was too sick to go in a carrying case underneath my seat. I had to monitor him the whole time, giving him water and food.

My dad, who lived in Florida at the time, met us at the airport and drove us to Greg's office. We had to wait quite a while, and we couldn't feed Alex because of the impending surgery. Alex was getting hungry. We sat in the waiting room, with Alex more and more urgently saying,

"Want banana," "Want corn," "Want water." I told him he couldn't, that he'd have to wait. Alex looked at my dad and said, "Wanna go shoulder." So Alex moved onto my dad's shoulder, cuddling up to his head, mumbling that he wanted this or that in a soft voice. My dad was a little deaf, so he didn't hear Alex's quiet entreaties. Finally, Alex screamed into my dad's ear, "Want kiwi!" Alex didn't even like kiwi, but he was obviously getting desperate. As tired and upset as I was, I couldn't help laughing.

I asked to be with Alex when Greg gave the anesthetic, so that Alex would be calmer and not require so much of the drug. I went back to the waiting room and fell instantly asleep—the first time in weeks, it felt like. An hour or so later Greg woke me and handed me a little bundle. It was Alex, wrapped in a towel. "He's going to be fine," Greg assured me. After a short while, Alex began to stir a little, and then some more. He opened an eye, blinked, and said in a tremulous voice, "Wanna go back." I said, "You'll be fine. I'll be back tomorrow."

I picked up a much chirpier, cheerier parrot the next morning, and we went back to Chicago. Alex had to stay at the vets' clinic for another couple of weeks for observation and to build his strength. We were also afraid that we couldn't entirely disinfect the lab of aspergillus before we left for Tucson, and we didn't want to risk exposing him again.

Alex became quite a fixture at the vets', talking to everyone who had time to stop and listen. His cage was right next to the accountant's desk. The night before I was due to take him to Tucson, the accountant had to stay late, working on the books. "You want a nut?" Alex asked her.

"No, Alex."

He persisted. "You want corn?"

"No, thank you, Alex, I don't want corn."

This went on for a little while, and the accountant did her best to ignore him. Finally, Alex apparently became exasperated and said in a petulant voice, "Well, what *do* you want?" The accountant cracked up laughing, and gave Alex the attention he was demanding.

That's my Alex.

Chapter 6

||||||||||||||||||||||||

Alex and Friends

Had things turned out differently, I would have contentedly stayed at Northwestern. I loved the gorgeous campus on Lake Michigan. I had some wonderful colleagues and close friends. The students there were terrific. And Alex was pushing the idea of what a bird brain was capable of achieving into territories no one had ever imagined possible.

But my wishes didn't matter. The academic authorities simply didn't know what to do with me. I was way outside the comfort zone of mainstream science, too edgy in the questions I was asking, impossible to slot into neat categories: neither psychology nor linguistics, neither anthropology nor animal behavior. I was part of all of these things but couldn't be shoehorned into any single one. So when I had been on campus for six-plus years, tenure not even a remote prospect, I had to leave. It was a rule, and they did know how to follow rules.

My marriage by this point had come to an impasse. It had been a very loving relationship to begin with, as marriages usually are. But David couldn't grasp the importance of what I was doing and why my work, often unsupported and challenged, should occupy as much of my time as his career occupied of his. I simply couldn't play a subordinate role. It was obviously best that we part.

So on the Sunday after Thanksgiving in 1990, Alex and I arrived at Chicago's O'Hare Airport, ready to check in at the United Airlines desk for a flight to Tucson, Arizona. I handed the agent two tickets. She was all smiles. She looked at them, looked around, and asked, "Where is Alex Pepperberg?" That was how Alex had been ticketed. I lifted the carrier so she could see Alex. He gave a cheery whistle. The agent instantly stopped smiling. "A parrot!" she blurted out. "Alex Pepperberg is a parrot?" she continued, putting a decidedly disparaging emphasis on the word "parrot." "I'm sorry, we don't sell tickets to pets," she said with great indignation.

"Actually, you do," I replied. "Here's the documentation." I showed her the sheaf of papers, pages and pages of it, that I had worked on with United's bureaucracy long before the trip: it confirmed that Alex was a valuable scientific resource (and somewhat of a TV celebrity) and thus needed a regular seat. It also certified that he was free of disease and could travel in the main cabin.

The agent would have none of it. She refused to listen to anything I said. I asked for a supervisor to intervene in what was rapidly developing into something of a Monty Python sketch. The supervisor saw that Alex did indeed have a legitimate ticket for a seat of his own. The agent, now chilly as a winter wind off Lake Michigan, begrudgingly checked us in. "What's that?" she then snapped, staring suspiciously at the three boxes I had at my feet.

"That's Alex's luggage," I said, by now highly amused at the unfolding farce. The boxes contained Alex's equipment, which I had carefully sterilized to make sure we weren't bringing any aspergillus with us. "I believe he's allowed three items, one to carry on, two to check," I said. "Isn't that right?"

The agent barely kept her anger under control. She tried for one last dig at me. "And I suppose you ordered him a meal?" she said with heavy sarcasm.

"Yes, as a matter of fact, I did," I said with a winsome smile. "He's getting the fruit plate."

When the fruit plate arrived, Alex turned his beak up

at it. He wanted my shrimp salad instead. The little guy knew how to travel!

It was with mixed feelings that Alex and I arrived in Tucson to begin the next chapter of our journey together. Yes, I finally had a tenure-track faculty appointment, an associate professorship in the Department of Ecology and Evolutionary Biology: the first "real job" of my unusual career, the first shot at security that tenure confers. As a woman, now on my own, that was important. But half the department faculty had opposed my appointment and had petitioned the dean to block it. Department heads at the University of Arizona have absolute power, and EEB's head, Conrad Istock, valued my work and believed my expertise would strengthen the department. The petition failed, and I was appointed. I had other supporters in the department beside Conrad, wonderful people, but the not-so-hidden antagonism made for an unsettling start.

Before very long I put that negativism out of my mind as I began to absorb Arizona's magic. I once wrote somewhere that Tucson brought tears to my eyes—literally, as I fairly quickly developed allergies to practically everything that grows there, but metaphorically, too, because of its beauty, majestic in its mountains, deserts, and giant saguaro cacti, and in its details, the animals, the smaller plants, and the birds. Oh, the birds! I had been an amateur birder ever since my dad put up the feeder in our yard in Queens. I became more serious as time went on. Now I had a birder's paradise on my doorstep, quite literally.

I bought a house about eight miles west of the city, an area that at the time was almost rural. Every morning I'd sit on my patio with a cup of coffee and gaze as the sun came up over the Rincon Range to the east. I watched as the sun's rays streaked across the highest peaks of the Santa Catalina Mountains directly in front of me, and sighed at the beauty of the lilac-pink haze that grew in extent and intensity over the Tucson Mountains to the west. That is the entire sweep of the Santa Cruz Basin, in which the city of Tucson nestles. And there it was for me to drink in its beauty every day. How could I not be entranced? How could I not be seduced by Mother Nature's splendor?

For the first time in my life I felt deeply connected to nature, the rich diversity of the Sonoran Desert fauna and flora right there in my own acre and a half for me to look at, smell, and touch. And in a part of the country where the Native American presence is so palpable, I was very much aware of that people's profound sense of oneness with nature. I resonated with that. All in all, I began to see Tucson as offering Alex and me great promise in our future work together, and as an opportunity for me for the first time to be my own self, not having to be something for someone else—a place and a time to restore my soul.

After a spell in temporary quarters, I established my lab in the basement of the Life Sciences West Building. Compared with everything I'd had before, this space was huge. There was a large central area, dominated by a round table that was Alex's domain (he also still had his ratty folding metal chair); there were two rectangular counters converging in one corner, one occupied by Alo, the other by Kyaaro, two young Greys I'd obtained from a breeder friend in southern California early in 1991 so I could expand our studies. (People had been constantly nagging me with comments like "But Alex is only one bird" and "What if Alex were to die?" The aspergillosis scare had made me pay attention to that latter possibility.) Each bird had his or her own room for sleeping, training, and testing. We had space for graduate students. And I had a huge office, truly palatial.

Before long I had four graduate students, which allowed me to expand the scope of study, adding behavioral ecology fieldwork in Africa to lab training and testing. I had an army of some twenty undergraduate students, who maintained the lab and entertained and trained the birds. To an outsider it must have seemed chaotic, and in a way it was. My philosophy was to have a culture that balanced playfulness and fun with serious, careful scientific study.

It was also very demanding, both on my time and financially. The department provided funds for one graduate student and National Science Foundation grant funds for another and for a few of the undergraduates. A bit more came from an undergraduate biological research program. But the rest came from The Alex Foundation, a nonprofit organization I established in 1991 to raise money to support all this work and for getting word of our discoveries to a wider audience. We ran fund-raisers and sold T-shirts and other Alex-related items, and I gave talks to bird societies. With all this activity and my new teaching responsibilities, I found myself working harder and longer hours than ever before. I never got back to my emotional oasis on the edge of the desert before ten-thirty each night and was rarely home on weekends.

Both Alo and Kyo, as we usually called Kyaaro, were sweet birds, but they came with baggage unknown to us or my breeder friend, Madonna LaPell. Alo had been mishandled by an owner before being sent back to Madonna. Alo seemed fine when I got her, at the age of seven months. She bonded with the students who worked with her, and we made reasonable progress. But when her caregiver students started to graduate and leave the lab, the trauma of her chickhood abuse kicked in. She must have felt abandoned, and began to screech pitifully when new people came near her.

Kyo, in contrast, appeared fine. True, he played really hard with his toys and had a shorter attention span than Alex had had when young, but then Kyo was only three months old. I didn't know what was normal. When he reached sexual maturity we realized that he had an avian version of attention deficit hyperactivity disorder, ADHD. He became harder to work with and would be startled by every small sound in the lab, running under a desk if someone were to drop a book or even a spoon. He couldn't seem to separate what was important in his environment from what was not.

Before these various problems began to manifest, however, we were able to embark on what was to be a long-term study of training methods. We wanted to know whether the model/rival method, which always involved two trainers and was time-consuming, was completely necessary. Could birds learn labels as efficiently with more streamlined approaches, ones that demanded fewer trainers and therefore less social interaction? We used audio and video tapes as exemplars, for example. The answer became very clear: the model/rival technique was more effective by far than anything else we tried. My early gut instinct and common sense had been supported: a rich social context is essential to teaching communication skills. I was tempted to say, *Duh!*

Alex, meanwhile, took quite a long time to fully recover from his near-death encounter from aspergillosis. One couldn't tell simply by looking at him, his behavior and comportment, but he was not fully healthy for at least a year after we got to Tucson. Greys, like most parrots, hide symptoms of disability, because in the wild, displays of vulnerability cause an individual to be shunned, as they are likely to attract predators.

For much of his time, Alex worked at being the boss of the lab, greeting visitors and directing activities from his round table near the middle of the room. We should have called him "Sir Alex," and some students began to call him "Mr. A." He liked to butt into Alo and Kyo's training sessions when he could. They usually were in their rooms during these periods, but on occasion we'd do a bit of review in the main room; then Alex would shout out a correct answer to a question while Alo or Kyo was struggling. Or he'd admonish them, "You're wrong," when, as too often was the case, they were.

I continued doing number work with Alex, including a study on recognizing and understanding Arabic numerals. That was a long, challenging project that was to come to spectacular fruition years later. I was working on a simpler number concept one day in the fall of 1992 while Linda Schinke-Llano was visiting from Illinois. I was showing Alex a tray of objects of different materials and different colors. "How many green wool?" I asked Alex. We had been doing this task for a short while, and Alex up to that point had been performing as he would for a TV crew, that is, spot on.

Alex eyed the tray and looked at me in a way he sometimes did, which I can describe only as wryly. "One," he piped. The answer was "two."

"No, Alex. How many green wool?"

That look again. "Four," he said in his charming singsong, double-syllable way: "Foo-wah."

Linda was watching, and I wanted her to see how well Alex was doing since his illness and rehab. "C'mon, Alex. How many green wool?"

It was no use. He kept alternating: "One . . . four . . . one . . . four."

By now I had realized he was just messing with my head. I knew that he knew the correct answer. "OK, Alex," I said sternly. "You are just going to have to take a time-out." I took him to his room and closed the door.

"Two . . . two . . . two . . . I'm sorry . . . come here!" Linda and I immediately heard coming from behind Alex's closed door. "Two . . . come here . . . two." Linda and I were laughing to the point of tears.

"I guess Alex is fully himself again," I finally was able to say to Linda. "The little rascal!"

Early in May 1992, I received a letter from Howard Rosen, a lawyer in Los Angeles. He wondered if he could visit Alex and bring his girlfriend, Linda. I often received such requests, usually from parrot people who have heard a lot about Alex through the grapevine. Mostly I politely decline, partly for reasons of security in the lab, partly to ensure the parrots' health, but also because my work schedule was extremely full. Howard's letter began, "Dear Dr. Pepperberg: This letter isn't coming from a nut case. Please don't discount it." That kind of statement is usually a sure indication that the letter is indeed from a nut case, and normally I would have discarded it.

But Howard explained in his letter that he was planning to propose marriage to Linda. Could Alex be trained to pop the question to Linda on his behalf—a kind of parrot proxy proposal? I wrote back and explained that Alex's language training didn't work quite like that. But I was so touched by this man's devotion to his girlfriend and by his inventiveness that I said, "Come anyway."

When Linda and Howard visited the lab very shortly thereafter, I heard the whole story. Linda was crazy about animals and had assiduously followed Alex's story in magazines and on television. She had taped a TV show of Alex and me and had shown it to Howard. He surreptitiously took the tape so he could locate me. He had planned to take Linda for a long weekend in Tucson and then spring the Alex wedding proposal on her as a surprise.

When Howard got my reply, he changed plans. He went out, bought a diamond ring and two plane tickets for Tucson, and made a booking at the Westwood Brook Resort, in the foothills of the Santa Catalina Mountains, just above Tucson. On the afternoon of May 8, he had Linda sit down; he got on one knee, as is proper. He then proposed marriage, proffered the ring and the plane tickets at the same time, and explained about the planned visit to my lab. Linda was ecstatic, as any woman in love would be at such a moment, with a proposal of marriage being made to her. "I'm going to meet Alex," she exclaimed. "That's just wonderful!" Howard told me jokingly that he was just a little downcast that Linda's response wasn't

"I'm going to be married. That's just wonderful." Such was Alex's celebrity. Howard and Linda flew to Tucson that very same day.

Alex wore his celebrity mantle with great ease and appeared to revel in the attention it brought him. Television crews from the United States and elsewhere came to the lab ever more frequently. Each time was another opportunity for Alex to show off, to flaunt his skills, to be the center of attention. He'd get a certain light in his eyes, metaphorically puff himself up, and step into the role of star of the moment. The public exposure earned us a good deal of notoriety. As so often happens in academic environments, it also stirred a bit of jealousy among my colleagues in the department, although I wasn't aware of it at the time.

Celebrity status was bestowed on Alex by adoring followers, too. Carol Samuelson-Woodson, an acquaintance, volunteered to help me take care of Alex over a Thanksgiving toward the end of my stint at Tucson. This was to be Carol's first encounter with Alex. Carol wrote about her experience in some detail in a charming essay. She described going through the security door into the lab and stepping in the disinfectant trays to prevent infection. "I finally came to a large, cluttered room where three beautiful African Greys gave me that inscrutable once-over," she wrote. "The closest bird played on a large, round table covered with what looked like the contents of an over-turned wastebasket: piles of shredded paper and globs of corn, berries, mashed veggies—very colorful. The other two parrots were on the counter, separated from each other."

The student in charge of the lab that day was telling Carol about schedules, food preparation, and so on. Finally Carol, who admitted to being very nervous, got up the courage to ask, "Is . . . one of these birds . . . Alex?"

The student said casually, "Oh, yeah, *this* is Alex," pointing to the bird on the round table right in front of Carol.

Carol wrote, "Stunned, I sank to my knees, rested my arm on the table to steady myself . . . I couldn't believe that here, right before me, was the fabulous creature, feathers and all, and that I had started off by SNUBBING him. 'THIS is ALEX?' I mumbled stupidly. I guess I'd expected a red velvet carpet leading to a magnificent throne; a golden perch and an impossible haughty bird in purple cloak and bejeweled crown. Then His Cuteness stepped onto my hand and marched right up to my shoulder, and basked in my belated adulation."

Alex never did sport the accoutrements of nobility, avian or otherwise. But Carol was right about the haughty aspect of Alex. Not all the time, and not always impossibly, but he filled that role with great aplomb when he felt like it. Which was quite often.

Ever since I started working with Alex in the mid-1970s, I had focused on his production and comprehension of labels and how he responded to requests or made his own—in other words, two-way communication between us—using parts of human speech. One of my first graduate students, Dianne Patterson, had a background in linguistics. This provided a wonderful opportunity to ask different kinds of questions about Alex's vocalizations.

I chose Alex as my partner in our long research project because I knew that Greys can produce English speech very clearly compared with other species of parrot. There are endless stories among Grey owners about the ability of their birds not only to speak clearly but also to sound almost identical to their owners. My friends Deborah and Michael Smith have a Grey, Charlie Parker. Charlie bonded strongly with Michael and speaks just like him, sometimes to Michael's embarrassment, other times to his advantage. Here's an example.

Debbie told me of a time when she was on the phone to a particularly obnoxious insurance agent. "He was being especially rude and overbearing, very loud," Debbie told me, "and I wasn't handling the situation very well." Charlie was getting more and more agitated by this verbal abuse, clinging to the side of his cage. "Charlie suddenly yelled out in Michael's voice, 'I'm going to kick your ass, you son of a bitch.' The guy was stunned into silence. I said to him, 'Well, I don't think we have anything further to discuss.' And that was the end of that ugly episode."

I don't have any stories to match Debbie and Michael's, but I am all too aware that Alex picked up the Boston accent that I acquired while living in the area, fairly mild though it is. When Alex said, "Want shower," it came out, "Want shou-wah," in the "charming" way Bostonians have of swallowing their *r*'s.

Obviously, my students and I have no problem understanding the sounds Alex makes. And our studies have shown that he has no problem understanding what we say to him. Dianne and I asked two questions. First, Alex's vocalizations sound like English speech to our ears, but are they really like English in their acoustic properties? Philip Lieberman, an eminent linguist, suggested years ago that parrots produce human speech by a clever combination of whistles, not the way you and I make word sounds. As a result, the acoustic properties of human and parrot sounds should be very different. Our second question was, how does Alex make sounds that to our ears resemble words when his anatomy is so different from ours in his vocal tract, his tongue, his possession of a beak, and, of course, the absence of lips?

To answer these questions, we used special equipment to record and analyze the sounds he made when answering our questions. And we installed him inside a heart X-ray machine so we could watch how his anatomical parts moved, or did not move, when he vocalized. I won't go into any detail here, because linguistic analysis can sound pretty daunting, but will focus on just one thing. A fundamental acoustic component of human speech is a so-called formant, an energy pattern that is characteristic of each speech sound we make. When a linguist looks at a sonogram of someone talking, for instance, she would be able to recognize what parts of speech—"oh" or "ee" or "ah," for example—are being produced. She would also identify the sounds as human.

When Dianne and I looked at sonograms of Alex's vocalizations, they looked, if not identical to what I would produce, then very, very similar, formants and all. Human speech, it turns out, is not as unique as it has long been held to be. Alex produces sounds that, acoustically, are very similar to the ones that you and I produce. No wonder we can understand each other, or at least hear what the other is saying. How exactly he does it is quite complicated and not really as interesting as the fact that he can. (Those of you who would like to know more will find what you need in chapter 16 of my book *The Alex Studies*.)

Our most exciting discovery was the way in which Alex produces the sounds we recognize as labels. Take "corn" and "key," for example. If Alex were simply mimicking these words, he would learn them and produce them as a complete sound. When you and I say "corn" and "key," however, we break the words up, so that the "kuh" in "corn" is different from the "kuh" in "key," and each word is completed with the appropriate sound, "orn" and "ey." The fancy term for this behavior is "anticipatory co-articulation." No animal was supposed to be able to do that. But, as Dianne and I demonstrated, Alex did that. Which means that such speech patterns are not as uniquely human as many people may think.

This behavior is one of a number of building blocks of language abilities, and here we find it in Alex's brain. Once again I need to stress that this does not mean that what Alex had was language, but that what he did makes us question a little more the nature of language and how it came to be what it is in you and me. On Alex's part, he once again demonstrated abilities he was not supposed to have.

Smart bird.

. . .

By the spring of 1995 I realized that pursuing long-term work with Alo was not feasible. Reluctantly, I sent Alo to Salt Lake City, Utah, to live with my friend Debbie Schluter, who I knew would give the bird the needed nurturing. Now we had to find a replacement. Branson Ritchie, one of the country's top avian veterinarians, told me that a friend of his, Terry Clyne, had offered to donate a Grey to my project.

Terry was a lawyer in Georgia, but her passion and avocation was breeding Greys at the Apalachee River Aviary. We talked on the phone. She told me she had the perfect candidate, a thirteen-week-old bird that was fully fledged and fully weaned, ready to go. We agreed that I would fly down to pick up the bird the following week, as I was going to be in Washington, D.C., and a side trip would be easy. This was early June.

I arrived at Terry's beautiful spread in Farmington, south of Athens, and she immediately took me to the breeding facility. Soon I was sitting on the floor amid a small flock of young Greys. It was quite a sight—and sound! Despite the fact that Terry had picked out a bird for me, she put me on the floor with all the babies she had at the time. Immediately I heard a *cheep, cheep, cheep, cheep* as the smallest of the flock tumbled over itself to reach me, barely able to toddle coherently, more quills than feathers on its tiny body. It started to pull at my jeans, and cheeped some more. He could not have been cuter, this fuzzy little thing with a disproportionately big head, eyes, and beak. A tiny bundle of energy and enthusiasm, all directed at me. Terry looked at me and said, "Well, Irene, I think, er . . ."

I nodded and said, "Yes, Terry, I think so, too." The little seven-and-a-half-week-old had chosen me. I simply could not resist.

After we stopped laughing and cooing, Terry asked me, "What do you know about hand-feeding Greys, Irene?"

I had never had to do it much; Kyo had needed only a bit of supplementary feeding, and that with a spoon. "Nothing," I said.

"Then we need to teach you how, fast," said Terry.

Greys as young as this little critter have to be fed using a syringe. It is very tricky, and potentially lethal. If you squirt the baby formula into the trachea rather than the esophagus, the bird dies. I got a one-hour crash course in how not to kill my precious new Grey.

"I want you to have this, too," Terry said to me as I was leaving, carefully carrying my infant Grey in a cat carrier. She handed me a small cardboard box. I opened it and saw a pink glass box in which, cradled in cotton wool, were fragments of a white egg, about an inch big, from which my chick had hatched less than two months earlier.

"Thank you, Terry," I said, and slipped the box into my purse.

When I got to the airport, I phoned Debbie in Salt Lake City and said, "Debbie, remember when you said I should call if there's anything I ever needed? Well, I need you to fly to Tucson to meet me. Now." I explained the unexpected situation to her. Debbie is a veterinary technician and skilled at feeding baby birds. We rendezvoused at the airport and then drove our little bundle across town to the lab, where Debbie spent several days teaching a team of us the skills, and hazards, of hand-feeding. There followed a nerve-wracking few months. Every day I lived in dread of the phone ringing in the morning before I left for the lab, bringing the news that something had gone terribly wrong. Thankfully, it never did.

Raising this little bird was like no other bird experience I had ever had. All the parakeets I'd owned as a kid had been fledged and weaned when I got them. Alex had been a year old when I got him, and Alo and Kyo, though younger, were still fully grown in size, if not yet sexually mature. All three Greys had for the most part been able to take care of themselves, feeding and preening. Not only did we have to hand-feed our new charge several times a day, but we also had to carry him around with us most of the time, wrapped in small blankets, for many weeks. The air-conditioning made the lab too cold for a bird with few feathers. And he would cry in distress if left by himself. He had been used to being cuddled with flockmates for comfort, hearing their heartbeats, feeling their warmth. He needed to hear our heartbeats when we held him. He needed our warmth. I had to be a parent substitute to him, preening his feathers and popping his quills (removing the keratin sheath from new feathers), because he had no mom to do that for him.

Inevitably, you bond in a very special way when you share these intimate times with a creature that is totally dependent on you. And so very cute. Alex would always own the number one spot in my heart—that went without saying. But this little newcomer nestled his way into there, too.

What to call the new bird? We settled soon on Griffin, for several reasons. First, it was in recognition of Donald Griffin, who had helped establish animal thinking as a legitimate field of science in the 1970s and 1980s. He had also helped get research funds for me while I was at Purdue. Second, some of us thought that his exaggerated chickhood features made him look like a gryphon, the fierce creature of myth who was half lion, half eagle. Last, *Griffin and Sabine,* a love story involving a parrot image, was a hot read in the lab that summer. So Griffin he became. (Because of the close, parent-substitute relationship I developed with Griffin, I ensured I would not take part in his training and would test him only in company with others. I felt I had to maintain that distance.)

A short time after Griffin arrived, we decided we would introduce him to Alex. When confronted with a youngster, adult birds may develop a caring, parental response, becoming protective. Part of our plan for future work was to have Alex play the role of a trainer for Griffin, partnering with one of us in the model/rival program. A good relationship between the two birds would therefore be very helpful. I carried Griffin over to Alex's table. Alex was busy working on his cardboard box, creating doors and other openings he could walk through, the way he would excavate a nest hole in the wild. I figured that introducing a new chick into this nestlike situation would be really easy.

I put Griffin gently onto the table. Alex stopped what he was doing, looked at Griffin, immediately growled his don't-mess-with-me signal, and began to walk slowly toward Griffin, feathers raised and beak poised menacingly. His intent was unmistakable. He was going for the jugular. I quickly snatched poor Griff out of harm's way, thinking maybe we should have put Alex in Griffin's spot in the lab instead of invading Alex's "territory." But it was now too late, and probably too dangerous, to try again. After the incident, Alex sat at the center of his table, the center of his domain, preening and wearing a rather self-satisfied look. We would just have to get along without a parental, caring Alex taking Griffin under his wing.

Territoriality is natural in Greys, especially in dominant birds, as Alex unquestionably was. Neither Alo nor Kyo was welcome on Alex's table, for instance; otherwise beak wrestling would break out if one of us didn't intervene in time. Also not welcome, it turned out, was a toy that looked and sounded like a Grey. In the mid-1990s there was a brief fad for toys that would repeat the last few words or sounds you made to them. A graduate student brought a stuffed-parrot version of one of these toys into the lab. She put it on Alex's table, and he adopted exactly the same stance he'd taken with poor Griffin. He approached the toy slowly, head thrust forward, beak out, producing that characteristic growling sound. Of course, the toy just growled right back. That infuriated Alex even more, and he seemed ready to tear the toy to pieces. It was removed to safety, and it never showed its beak in the lab again.

Not all toy birds provoked this aggression, it turned

out. Some must trigger a primal response, while others do not. After a local television program had featured Alex, someone sent him a toy parrot, one that played songs when you pushed a button. We suspended it over one side of Alex's table, and he completely ignored it.

After about a week, one day he looked intently at the suspended parrot, walked up to it, and said, "You tickle." He then bent his head over toward the toy, the way he would to a student, who would then dutifully tickle Alex's neck. Nothing happened, of course. After a few seconds he looked up at the toy, said, "You turkey," and stalked off in a huff. The students had sometimes said, "You turkey" to Alex when he did dumb things. He had apparently learned how to use that stinging epithet without any training.

Bernd Heinrich was a professor of zoology at the University of Vermont, now retired. One of his passions is crows and ravens. Heinrich and I share the same curiosity about bird intelligence. One day in late 1990 he decided to test the widespread belief that ravens are especially smart. Heinrich tied a piece of meat to the end of a piece of string some thirty inches long. He attached the other end to a horizontal branch of a tree in his home aviary and sat back to see if the ravens could figure out how to get access to the meat. (It was dried, so they couldn't just break off a piece and fly away.)

After a while, one of the ravens landed on the branch next to the string, leaned down, took the string in its beak, hoisted it up, and anchored the resulting loop between one of its claws and the branch. The bird did that half a dozen times, until it had brought the suspended meat within reach. It looked to Heinrich as if the bird had assessed the situation, worked out a plan for retrieving the meat, and put it into action. No trial and error, no practicing. The bird never attempted to fly off with its prize, even when Heinrich shooed it away. Apparently it understood that the meat was securely tethered.

This seemed to me to be an elegant exercise, and I have to admit to a little edge of competitiveness on the bird brain front: if Heinrich's ravens are smart enough to do it, what about my Grey parrots? Shortly after Heinrich's paper came out, in 1995, I set up a similar challenge in the lab. I used a favorite bell rather than meat, given a Grey's taste. I put Kyo on the perch; he looked down at the suspended bell, and then did exactly what the raven had done, using beak and claw to gradually haul up the bell. Score one for Greys.

Then it was Alex's turn. For him, I used an almond,

something he liked more than any toy. I put him on the perch. He looked down at the nut and looked at me. He did nothing. I was wondering what he was thinking. After a few seconds he said, "Pick up nut."

I was a bit taken aback. I said, "No, Alex, *you* pick up nut."

He looked right back at me and said, "Pick up nut!" a little more insistently this time.

I tried to encourage him several more times, but he simply refused. With one bird doing the task as planned but Alex refusing, we didn't try to publish the results, and it was only several years later, with Griffin repeating Alex's actions and yet another new, not-yet-really-talking bird repeating Kyo's actions, that I realized what was happening.

Once Alex had learned how to label objects and request things, he relished the control it gave him over his environment, the ability to manipulate the people around him. Alex's boss-of-the-lab personality had emerged during our years at Northwestern. By the time we had settled in at Tucson, it was in full bore. The students used to joke that they were "Alex's slaves," because he would have them running around, attending to his constant demands. He was merciless with new students. He would run through his entire repertoire of labels and requests: "Want corn

. . . want nut . . . wanna go shoulder . . . wanna go gym," and on and on. It was Alex's initiation rite for newcomers. The poor student would have to run around and fulfill all these wishes, otherwise he or she would never get anywhere working with Alex.

Alex's "failure" on the string-pulling test was not a black mark on his intelligence, I realized. It was a measure of his sense of entitlement, his expectation that I would do as he asked. If I were going to do something as silly as hanging a nut on a piece of string rather than handing it directly to him, as was usual, then I would have to give it to him when he asked for it. Otherwise he would have none of this game. Why had Kyo succeeded where Alex had not? Probably because at the time of the experiment, Kyo had little command of labels and requesting, and was thus much less used to getting people to do things for him. He relied on his native Grey intelligence to get what he wanted. Alex, in contrast, relied on his entitlement.

The days were full for the birds, with training sessions, sometimes testing, often being entertained by the students, or, in Alex's case, ordering them around. Come five o'clock, the students left for the day, and then it was just me and the birds, hanging out. Kyo was less sociable and preferred to go to his cage at this point. I then had dinner, with Alex and Griffin as company. Dining company, really, because they insisted on sharing my food. They loved green beans and broccoli. My job was to make sure it was equal shares, otherwise there would be loud complaints. "Green bean," Alex would yell if he thought Griffin had had one too many. Same with Griffin.

Later in their relationship they developed a comical little duet: "Green," Alex would pipe up.

"Bean," Griffin responded.

"Green."

"Bean."

"Green."

"Bean." They would go on like that, alternating, with ever more gusto.

After dinner I took them to my office and placed them on their respective perches, so they could watch me do e-mail and work at my computer. They would constantly ask for treats, such as nuts, corn, and pasta. Alex's perch always had to be a little higher than Griffin's, as he was "senior bird." Wherever we were, Alex had to be top bird, quite literally. Alex always remained jealous of Griffin, perhaps because of the attention we gave Griffin when he was a chick. Whatever the reason, if I came into the lab and said hello to Griffin before acknowledging Alex, I could forget about working with Alex. He would sulk the whole day.

Our plans to have Alex act as a tutor to Griffin worked out to a degree. But Griff always learned more efficiently when he had two human tutors rather than one of us and Alex. We aren't exactly sure why. There are several possibilities. One is that Alex always treated poor Griffin as if he were a pain in the butt, and perhaps Griffin felt inhibited by that. And at the time, Alex wouldn't question Griffin, so we couldn't exchange roles of model/rival and trainer, a crucial part of the procedure. Or maybe Griffin figured that the exchanges between Alex and the students were special pair-bonding duets in which he wasn't supposed to engage; Greys in the wild do have such duets with their mates.

Also, Alex could often not resist showing off. He'd sometimes give the right answer when Griffin hesitated. Or he'd tell Griffin, "Say better," which meant Griffin should speak more clearly. Alex also occasionally gave wrong answers, apparently to confuse Griffin. Griffin was always good-natured and put up with Alex's antics and high-handedness.

Alex was happy in the lab, as were the other birds. And why not? They enjoyed far more attention than the vast majority of pet birds. But from time to time I took Alex home with me, to give him a change of scenery. He loved to sit by the window and see the trees and sun himself. It wasn't always easy having him at the house, because he wanted me constantly with him. He hated to be put in his cage during the daytime, if I needed to do an errand, for instance. But as long as I was there and he was free to sit near me, he couldn't be happier.

All that changed one day in 1998. I had just brought him into the house and set him on a perch when he became terribly distressed, squawking and saying, "Wanna go back . . . wanna go back!"

I rushed to him and asked, "What's wrong, Alex? What's wrong?"

I looked out of the window and quickly realized what alarmed him. A pair of western screech owls were building a nest in the roof over the patio. They apparently struck terror into poor Alex, even though he had never seen an owl in his entire life. I tried to calm him, with little success. I pulled the drapes, so he could no longer see the owls. Still no use.

"Wanna go back . . . wanna go back!"

It was a great demonstration of object permanence. Even though Alex could no longer see the owls, he knew they were still there. And even though they were outside the house and he was safely inside, he was still terrified.

Reluctantly and sadly, I packed him into his cage and drove him back to the lab that evening. I knew he would never return, that this was the last time in my house for Alex and me. I saw, too, that even though he had lived his whole life in the company of humans, even though he had lived all but one of his years with me, even though I thought of him as my Alex, there was, and always would be, something in him beyond the reach of any person, even me. When the image of that little owl entered Alex's brain, a brain innocent of any such images, it triggered an urgent and instinctive message: *Predator, danger, hide!* It was a primal response, something imprinted in his DNA.

And I could not calm him.

Chapter 7

||||||||||||||||||||||

Alex Goes Hi-Tech

My students and I spent countless hours with Alex, teaching him to produce and comprehend labels for objects and concepts. His achievements were impressive. But often it was the labels and phrases he picked up in passing that were especially memorable. So it was when Alex told me one day to "calm down."

As the 1990s drew to a close, my work environment became more and more stressful. Although I had been granted tenure not long after moving to Tucson, I was still an associate professor. In 1996, I came up for promotion to full professor, which was denied. Although it was never stated explicitly, my oddball circumstance—a chemist in a biology department—did not help my case, I'm sure.

What was explicit was increasing pressure on me to teach introductory biology. I felt that was entirely inappropriate for someone with my background. I considered the courses I did teach—animal-human communication, for example—to be valuable contributions to the department of a major university. Instead, they were dismissed as "designer courses," not germane to producing the maximum number of graduates each year.

Also resented was my public exposure, with Alex starring in so many television and print stories. Jealousy is corrosive. When the time came for my sabbatical in 1997, the year following my nonpromotion, I grasped it eagerly. I had won a Guggenheim Fellowship that would enable me to write up my two decades of work with Alex, in *The Alex Studies*, a book with Harvard University Press. And it gave me a break from the day-to-day rancor. Yet it probably caused even more ill will. I was asked to put the sabbatical aside and teach intro biology; again I declined.

As Tolstoy might say, every unhappy workplace is unhappy in its own ways, but the pattern is the same: some mixes of people, rules, and settings just don't add up to a positive result. I won't bore you with the details, just one funny Alex story. After one meeting in early fall 1998, after I returned from my sabbatical, I was irked more than usual. Exactly why, I cannot recall. In any case, I was fuming when I left the meeting, cursing my fate at being stuck where I was, with no prospect of a way out.

Usually when I walked down the corridor to the lab I would hear a cheery whistle from Alex, his greeting to me. He had grown familiar with the sound of my footsteps on the tiled floor, and that was his signal to begin his greeting. On this occasion, however, there was no whistle. I flung the door open and stormed into the lab.

Alex looked at me and said, "Calm down!" He must have heard something distinctive in my footsteps that alerted him to my emotional state. I stopped in my tracks when I heard him say that. Had I not been so annoyed, I might have said something like, "Wow, guys, did you hear what Alex just said?" But I didn't. Instead I looked directly at Alex and snapped, "Don't tell *me* to calm down!" I slunk into my office.

About a year later, that little interchange was printed in the *New York Times* as its quote of the day. A reporter for the *Times* had included the quote in a story about Alex and me. "Sometimes," she said in the story, "Dr. Pepperberg and Alex squabble like an old married couple."

About a month later, I received an e-mail out of the blue from Michael Bove, head of the Consumer Electronics Laboratory at MIT's Media Lab. Would I like to give a lecture at the Lab on my work with Alex? Conceived in 1980 by architecture professor Nicholas Negroponte and former MIT president Jerome Wiesner, the Media Lab had become one of the most celebrated research institutions in the United States, at least according to the popular press. It had a reputation for supporting brilliant, wacky techno-geeks whose mission was "inventing the future," as Stewart Brand put it in his 1987 book on the Lab. It defined "cool" in the related worlds of technology and communications.

So I was aware of the place from what I'd read in newspaper and magazine articles, although I could not imagine why they would want to hear from a woman who talks to parrots. Nevertheless, I said yes. At the very least I would visit my beloved Boston.

The Media Lab is housed in a suitably futuristic, white-tiled building on Ames Street in Cambridge, known locally as "Pei's Toilet" after its famous architect, I. M. Pei. I arrived in early December and was met by Mike Bove. Mike offered to show me around before we had lunch, and he warned me, "The Media Lab is very often a bit of a shocking place for people who have never been here before." He was right. You emerge from elevators on the third floor to be confronted by a teenage boy's wildest dream of technology heaven. It's all glass walls and computers and, in the Lab's lingo, "stuff" everywhere: on the floors, on the walls, hanging from the ceiling. Mike told me that at night the corridors are haunted with crazy things like mini-robots and odd automatons roaming around.

The free spirit of the place encourages innovation and rebelliousness as the norm. Being on the edge was not just allowed but expected. Different sections of the third floor were known variously as the Garden, the Jungle, and the Pond, the last referring to the primordial ooze from which all kinds of unexpected creatures might emerge. Nothing I had read or seen prepared me for how far outside the normal bounds of doing and thinking the Lab really was. I was entranced.

Over lunch, Mike casually said, "Have you ever thought about spending a year on leave here?" I was stunned. Such a thought had never even crossed my mind. But I knew instantly that this was where I wanted to be more than anything. I said something like, "Oh. How fast do you want me here?" and "If you want me permanently, I can arrange that, too." I was ecstatic. Here was an opportunity to move beyond the rut in which I had found myself in Tucson. I felt I would be coming home geographically, to Boston, and intellectually, in the way I liked to function as a scientist. At least I'd have some fun for a year. Who knew what might lie beyond that?

· · ·

There was only one problem. Although I would be able to work on my parrot studies, I didn't feel right about moving Alex across the country twice in one year, so I made the painful decision to leave him in the capable hands of my lab colleagues and students and endure a year's separation. When I drove out of Tucson for Boston the following August, it was with very mixed feelings. Ahead lay the promise of a year of exciting science of a kind I had never dreamt: applying cutting-edge technology to cognitive questions I had thought about for so long, and in the company of freewheeling thinkers. What more could I have asked for, a permanent job aside? Well, Alex. I arranged to spend one week in four back in Tucson, keeping the lab going, advising the students, and seeing Alex and the other birds. I had always traveled a good deal through the years, so being separated from Alex for a few days wasn't new. But this was serious separation. It would be hard, for me and for Alex.

One of my first tasks at MIT was to find a Grey for my new work there. I had spent much of the previous six months thinking about what I might do at the Lab. The reason I had been invited was a shared interest in intelligent learning systems. Several people there were studying computers as learning systems, and hoped to use my parrot studies as models. We could learn from each other in this realm. But I saw other possibilities. One of my long-standing concerns was finding ways to keep pet parrots from getting bored, which is a common problem. They are highly social, highly intelligent creatures. If they don't have constant attention and things to do, they become distressed, sometimes to the point of being psychotic. They may screech and pluck their feathers. Many people who get Greys and other parrots as pets don't appreciate this about their birds. It is quite cruel to leave them in a cage by themselves all day long. I've given many a talk at bird clubs about this issue.

Given the Media Lab's resources, I wondered if I could use technology to keep domestic parrots entertained and happy. I needed a Grey that was old enough to manipulate simple equipment. A baby bird would not do. I got Wart, a one-year-old Grey, from Kim Gaudette, a parrot breeder in Connecticut. Actually, we named him Arthur, but he was immediately known as Wart, the name Merlin the magician gave the young King Arthur. Poor Wart had a damaged foot from an accident as chick. He could perch and pick up food, but he wasn't as stable as other Greys. We didn't clip his wings as much as usual, so he could save himself if he fell off a perch. This led to more than one occasion when Wart took off and flew loops around the Pond, causing great consternation and amusement. He also liked to visit his favorite secretary, who fed him forbidden potato chips and french fries.

The philosophy of the lab was simple: *there's a ton of resources here, physical and intellectual, so go do interesting stuff.* I soon teamed up informally with Bruce Blumberg, whose focus was on learning how dogs make decisions, and in building computer systems that learn the kinds of things that animals learn easily. He also had a dog, a silky terrier, a cute little thing that Bruce said was his inspiration. Our intellectual interests overlapped a lot, and because he was the "dog guy" and I was the "bird lady," we called ourselves the Woofers and Tweeters.

Before long, I was working on various projects with some of Bruce's graduate students, Ben Resner in particular. One project was Serial Tr-Hacking ("hacks" being MIT practical jokes based on technology), in which Wart had to identify instructions, in the form of simple images, and then perform a combination of pulling, flipping, and rotating a lever to have a food treat delivered to him. Another was an electronic bird sitter, designed to stop the screeching parrot problem. This involved a screen on which the parrot watched pictures and videos. Exactly what images were displayed depended on how much noise the bird was making. Below a desired level, the pictures would be positive images, such as the bird's owner, parrots in the wild, that kind of thing. If the parrot's whistles and screeches exceeded the desired level, the pictures would switch to negative objects, such as a raptor swooping or a ground predator creeping nearby. The idea was that a microphone system would track the sound level of the bird's calls and control the images accordingly.

We were also thinking about something we called a "smart nest," a device to track Greys' behavior at nest sites in Africa. We were working on global positioning system tags that were small enough and light enough to be attached to Greys' backs, so we would be able to monitor their movements during the day.

And we were developing a system that might be used both for enriching a parrot's vocabulary and for working with autistic children. We imagined having a series of "toys," each of which would have a radio frequency identification tag. The action of picking up a toy would trigger the playing of a video about the toy. For instance, if the toy was a key (parrots used keys for scratching themselves), then the video would be of a person saying something like, "A key. You've got a key. Wow! Look at the key." When the toy was put down, the video would stop. That was the idea, anyway. We called it our PollyGlot Computer. The

ALEX & ME

challenge was to have enough variation in the videos so that the subject wouldn't get bored.

If that one sounds a little sterile, it was. Boredom was a constant issue with Wart. It was hard to make these games interesting enough to keep his attention for very long. If an undergraduate entered the room during a session, Wart was likely to find the newcomer far more interesting than pulling or pushing a lever one more time. Ben used to say that he and his fellow students felt they were in competition with Wart: the students were trying to outsmart Wart, while Wart was busy outsmarting the students. The students told me they felt as if Wart were saying, *Hey, give me a real problem. This one is just too easy for a smart bird like me!*

Bruce and I put on a mini-symposium, called Wired Kingdom, the spring after I arrived. We had people describe all kinds of ways of using electronic gadgets to study animals in the wild and in zoos. I was having a blast, not least because of the opportunity to interact with such bright and motivated students. I loved that. Apparently the Lab's directors did, too, because shortly after our symposium they asked me to stay a second year, which was unheard of for a visiting professor, as I was. This time I was determined not to leave Alex in Tucson. He and Griffin were coming to Boston to be with me and Wart.

Wart was also having a blast. When I think about the birds' personalities, I always come to an amusing contrast. Alex was like the kid in class who always knows the answers and is constantly jiggling around in his seat, his hand waving high, wanting to be the one to be chosen to answer the teacher. Griffin is like the smart but shy kid, trying to make himself invisible so he won't be chosen. Wart is the party animal, the teenager playing hooky with his friends. He was also like the techno-freak teenager, perfect for his role at the Lab. He was good at manipulating the equipment, and he loved it. He was confident and at ease performing in front of an audience.

There were always audiences at the Media Lab. The Lab was funded by sponsors, corporations who got to see and benefit from what was going on in the Lab, for a sizeable fee. There were always visitors coming through, but the main events were two sponsor weeks, also known as "demo weeks," one in the spring and the other in the fall, very lavish affairs. During the time running up to demo week, the Lab was a frenzy of activity as the students put the final tweaks on their demos, practically living there. The Lab's saying, "Demo or die," was taken very seriously.

Those were intoxicating times to be at the Lab. The stock market was setting records every day. Dot-com mania was in full bloom. The Lab seemed awash in corporate money. When I learned early on that my NSF grant had not been renewed, the Lab essentially said, "Don't worry; just do your work." I thought, *Wow! This is* not *Tucson.*

There was never any shortage of cool things for sponsors to see during sponsor week, but a live parrot in a demo was always an added attraction. Wart performed wonderfully at our first such event, in the spring of 2000. Every fifteen minutes or so a group of people would come through my lab, and we would demonstrate our projects. Wart would pull and flip levers, as instructed. He was a natural.

By the end of the week, the poor bird was exhausted. It was the last day, and one more sponsor came through. Wart was sitting absolutely still on his perch, eyes closed, napping. The sponsor came into the lab, saw Wart, and stooped to take a closer look. Wart slowly opened an eye, then closed it. Otherwise he was motionless. The sponsor exclaimed, "Oh! Animatronics!" I said, "No, no. He's not a robot. He's a real bird. Don't get too close; you might get bitten!"

I was thrilled by the prospect of having Alex and Griffin with me again, and I looked forward to introducing them to this new world of technology. But there were hassles in getting them to Boston. I was finally able to bring them with me in separate carriers on a red-eye flight, connecting through Dallas. The poor birds were miserable from the time I put them in the carriers to when I let them out in Boston, some twelve hours later. They had refused to eat anything. I had taken them to the plane's bathroom with me, trying to coax them with treats in private. But they were too stressed. Alex was a particularly sorry sight, his tail feathers all but chewed off. It is common for parrots to overpreen their feathers when they are distressed, and the months of our separation had been hard on him. I had seen that on my periodic visits to Tucson. Each time I got back he struggled with the pleasure of seeing me again and the anger at my being away from him. But now we were together.

Shortly after Alex and Griffin arrived at the Media Lab, in September 2000, a producer for *Scientific American Frontiers*, a public television science show, contacted me about taping a segment of an episode. I had done one a decade earlier, in Tucson. The current show was to be on pets and technology, called "Pet Tech." Alan Alda was the presenter. I was going to meet Hawkeye Pierce!

When Rebecca, my unofficial godchild, heard about this, she was wild with excitement. She is a huge fan of *M*A*S*H*, and she begged me to get an autograph for her. I confess I was a little bit nervous before meeting Alda, unusual for me. But he also turned out to be a really nice guy, very amusing and friendly. When I proffered Rebecca's copy of the *M*A*S*H* book and asked if he would autograph it for my godchild, he raised an eyebrow, just the way I'd seen on screen. "Of course," he said. "And what is your . . . er . . . godchild's name?" I told him it was Rebecca. He laughed and then said, "So it really *is* for your godchild!" I frowned, puzzled by what he meant. "Well, people often *say* it's for their godchild, but it's really for themselves, and they are too embarrassed to admit it." He signed the book, and I gave him a copy of *The Alex Studies* in return.

It certainly was one of the most fun taping sessions I had done, partly because of who Alda is and how very charming he is, and also because he was so obviously enthralled with Alex and what he could do. The show started with a few clips from the previous show, in which Alex identified the color of an object, responded to "How many?" and "What color bigger?" The shot then cut to Alda, me, and Alex in the Lab. "Hello, Alex," he said. Then he turned to me and asked, "Can Alex do any new things since last time?"

I said, "Yes, he can," and proceeded to demonstrate.

I showed Alex two keys and asked, "What toy?" I then asked him, "How many?" and "What's different?" Alex was in good form and answered quickly and correctly, if a little mumbly on the final answer.

I then held up a tray of plastic Arabic numbers, each a different color. By this time, Alex had learned numbers up to six. "What number green?" I said.

Alex hesitated and said, "Want a nut."

I said, "Come on, Alex, you can have a nut later. What number is green?" I was thinking, *Oh, no. Is he going down that path?*

But then he quickly said, "Four," which was correct. Then he again said, "Want a nut." I gave him one.

Alda was sitting there shaking his head, grinning widely, clearly impressed by Alex's talents. I explained the model/rival technique for teaching Alex labels and concepts, and suggested we do a session right then and there. I asked Alda, "What toy?" holding up a spoon. We went back and forth on that.

Then Alda turned to Alex and said, "Alex, what toy?"

"Want a nut," said Alex.

Eventually Alex produced a "sss" sound, and then something that approximated the full word. It is a difficult word for him to say.

"I find this very hard to believe," said Alda, "that Alex is really doing what he seems to be doing." He then turned to the camera and said that he was well used to coming to the Media Lab to film interesting projects. "I have to say that Alex and the other birds look oddly out of place in an institution best known for high technology." He explained my pursuit of gadgets for entertaining parrots. "To find out why parrots need entertaining, we are going to go to Foster Parrots, about an hour south of Boston."

Foster Parrots is a parrot rescue facility run by Marc Johnson. There we had a conversation about the problems people have with pet parrots that are left alone all day, the kind of thing I had been proselytizing about for years. "It is like putting a four-year-old kid in a playpen in the morning and leaving him there all day by himself," I said. "Of course he would be angry and upset when you got home. It's the same with parrots."

"You have to remember that dogs have been domesticated for thousand of years," explained Johnson. "Mostly they are fine being in a home environment by themselves. But parrots are not domesticated. They are wild animals, and we shrink their world to the size of a single room in your house, or even to the size of a small cage. We have to remember that. That's why we need ways of entertaining them when you aren't there."

The show then cut back to the Lab, where Wart was expertly working with the Serial Tr-Hacking equipment. I then described what we called the InterPet Explorer. Ben Resner had suggested the idea at an early brainstorming session on finding ways for parrots to entertain themselves. "Why not find a way for parrots to be able to surf to sites they find interesting?" He meant it more as a joke than a serious idea, but the Media Lab people loved it, and we instantly got funds for developing it. When Ben finally had something up and running, it wasn't really a way for Alex and friends to surf the Web, though journalists loved that idea. Headline writers had a lot of fun: "Give That Bird a Mouse" was one example; "Polly Wanna Web-Surf" was another; "On the Internet, No One Knows If You're a Parrot" was yet another.

The idea ultimately came down to Alex, or whomever, being able to select between four modes using a joystick: pictures, music, games, and video. Within each mode were four choices. In music there were clips of a classical piece, a rock song, jazz, and country music. Wart came to be quite proficient at surfing the availabilities, as did Alex. But Griffin showed little interest. Our hope was that a bird alone at home would be able to entertain itself for hours with its favorite activities. One practical problem was to generate a large enough number of options. How many times would you want to listen to the same fifteen-second clip from Vivaldi's *Four Seasons* before you'd want something else?

At the time we shot the *Frontiers* show, the prototype InterPet Explorer had just been put together. Alex had had little experience with it. As Alda said, "During filming, Alex remained resolutely uninterested in the browser." That was no surprise to me. Alex was much more interested in Alda and me, and the rest of what was going on around him, than in this limited little machine. So we decided to film him in front of the browser when there was no one else in the room. He did pay attention to it, though he was not very interested in the pictures. Instead, he loved the music. The *Frontiers* show ended with Alex perched in front of the screen, repeatedly selecting the music option, and enthusiastically whistling along with whatever was being played, head bobbing, obviously having some private fun.

Everything about the Media Lab was such a great improvement over my situation at Tucson, except it was cramped. I had an office of my own, which I shared with Wart. He was regarded as something of a Lab pet. He was not alone in that role: there were several dogs wandering around.

Alex Goes Hi-Tech

Our work space, just off the third-floor Pond, measured about ten by fifteen feet—not a big room. Ben had a desk in there, as did Spencer Lynn. Spencer had been a grad student with me at Tucson, where he had been Alex's absolute favorite person. I had a special relationship with Alex, obviously, but in general Alex preferred guys, especially tallish guys with longish hair, like Spencer. Alex often would pad around the Tucson lab, looking for Spencer. When Spencer picked him up, Alex would run up his arm, perch on his shoulder, and perform the Grey's mating dance. Spencer was the only person Alex called by name. He used to say, "Come here, Ser."

In 1999, however, Spencer had committed a cardinal crime, in Alex's estimation, when he went to Africa for three months to study the behavioral ecology of Greys. He had abandoned Alex. And Alex never forgave him: Spencer was no longer his out-and-out favorite. And now they shared a small room at the Lab, together with Ben, Griffin and Wart, a temporary programmer, several computers, all kinds of other electronic equipment, soldering guns, two bird cages, and a troop of undergraduates whose job it was to train the birds. I don't know how Ben and Spencer ever got anything done. They would sit at their desks, trying to read, work on their computers, or construct something for one of the Pet Projects. Within a

The waiter came up to our table and said, "Would you like to hear about our specials?"

We said, "Sure."

He said, "We have Chilean sea bass in a pesto sauce, served with squash and green beans."

My wife and I looked at each other and did the "green" . . . "bean" duet that Alex and Griffin often did. The waiter gave us a Are these guys nuts or what? *kind of look.*

Mr. A. did have a way of getting into people's heads.

The fall 2000 sponsor event was scheduled for a few weeks after the *Frontiers* filming. I hadn't planned to have Alex demonstrate what we were doing with phonemes, the individual sounds that make up a complete word, but sponsors had asked to see that task. We had started this project in Tucson and continued it at the Media Lab. We were training Alex to sound out phonemes, but not because we wanted him to read as humans do. Instead, we wanted to see if he understood that his labels are made up of sounds that can be combined in different ways to make new labels. We knew that he sometimes babbled when alone, producing such strings as "green, cheen, bean, keen," and so on. This suggested that he did indeed understand that labels are made of subunits that can be used in different ways. But, as always, we needed more scientific proof.

We used plastic refrigerator letters, each a different color. We taught him the sounds of the different letters or letter combinations. We would ask him, for example, "What color is 'ch'?" and "What sound is purple?" He had become quite proficient.

We had a very short amount of time scheduled for the demo, and the sponsors were very keen to see Alex do his stuff. I showed Alex a tray of his letters. "Alex, what sound is blue?" I asked.

He answered, "Sss."

It was an *S*, so I said, "Good birdie."

He replied, "Want a nut."

Because we were pressed for time, I didn't want to waste it with Alex eating nuts. I told him he had to wait, and asked, "What sound is green?"

Alex answered "Ssshh."

Again he was right. Again I said, "Good parrot."

And again Alex said, "Want a nut."

"Alex, wait," I said. "What color is 'or'?"

"Orange."

"Good bird!"

"Want a *nut*." Alex was obviously getting more than a

little frustrated. He finally got very slitty-eyed, always a sign he was up to something. He looked at me and said slowly, "Want a nut. Nnn . . . uh . . . tuh."

I was stunned. It was as if he were saying, *Hey, stupid, do I have to spell it out for you?* More important, though, he had leaped over where we were with his training, which was individual phonemes, and gone on to sound out the parts of a complete word for us. Perhaps he was really saying to us, *I know where you're headed with this work! Let's go on with it. Let's do whole words!* It was a stunning moment, and it made me wonder just how far beyond our expectations Alex was going to lead us in the years ahead.

A few months later, the future looked even more promising when the possibility was raised that I might be able to stay at the Lab beyond the two years that had already been agreed. There was a lot of back and forth over whether I would be able to get a professorship or a long-term contract as a research scientist. It was a tense time for me, because I needed to know well before the beginning of fall term 2001, the time I was due to finish my leave of absence and return to Tucson. At one point in August I had two moving trucks reserved, one in Boston, in case I had to move back to Tucson, and one in Tucson, in case I was going to stay in Boston and needed my things there.

Finally, at the very last minute, I received a five-year renewable contract as a research scientist, with all the financial support I would need for my work. True, I was going to have to give up a tenured job for a nontenured one. But I could not have been happier. I wrote to Tucson and told them I would not be coming back. My future could not have been more promising. I would be able to continue the cognitive explorations I had pursued for so many years, and I was going to continue to explore technological applications for them. I would not have to worry about research funding. And Alex and I would be together.

Three months later, in mid-December 2001, I learned that I was one of thirty people whose jobs were to be eliminated at the Lab. There had been storm clouds gathering about the Lab's financial health. The technology-heavy NASDAQ Composite index had peaked a year earlier and had begun what was to be a catastrophic decline, marking the end of the dot-com bubble. Then 9/11 happened, which exacerbated the economy's woes. Corporate sponsors of the Lab were no longer able to keep up their previous level of support.

When I had arrived at the Lab two years earlier, it was at the peak of its technological and financial exuberance. I had imagined a future with limitless possibilities for my research. Now I had no job and nowhere to continue my work with Alex and friends.

Even before the December announcement, there had already been problems with where to house the birds. In September I had moved Alex and Griffin to Newton, a suburb of Boston, to live in Margo Cantor's house. Margo's son had been one of Alex's MIT trainers, and she kindly agreed to look out for the birds until we sorted out housing arrangements. Wart, meanwhile, went to live with my friend Maggie Wright, in New York City. These were meant to be temporary arrangements, lasting no more than a few weeks before we would move them back to new facilities. Now I had no idea where they might go, or when, or how we would support the research. Or how I would support myself.

Chapter 8

||||||||||||||||||||||

The Next Horizon

Alex was miserable and angry. Margo Cantor and her husband, Charlie, could not have been kinder in giving refuge to Alex and Griffin in their house in Newton. Alex really liked Charlie, and Griffin was very affectionate with Margo. But Margo and Charlie were gone all day long, leaving Alex and Griffin locked in their cages. It was just the situation I had always warned parrot owners to avoid. I spent my days at MIT working on manuscripts, applying for jobs, and trying to find lab space nearby to house the birds properly.

Each afternoon I drove the eight miles from Cambridge to Newton. I tried to greet the birds cheerily, though my mood was very dark. Alex often stuck his beak in the air and turned his back to me, punishing me for abandoning him. He sometimes refused to come out of his cage, quite uncharacteristic for him. I stayed with him and Griffin until Margo came home, around six. I then went back to the Lab to cram in a few more hours of work. Both Alex and Griffin were very subdued during this trying period. They showed their stress by plucking feathers as the temporary housing arrangement stretched from an anticipated few weeks to an eventual five months.

Even before I'd been let go from the Lab, I'd begun searching for lab space. The room the birds had shared with Ben and Spencer was needed for other projects. I was fortunate that Bob Sekuler, a friend and vision physiologist from Northwestern who was now at Brandeis University, had offered to help me secure space there. Brandeis University was relatively close by. I managed to obtain a room in the animal care facility of the psychology department. All it really needed was a coat of paint. It was mine pretty much as long as I could pay the rent.

Eventually, when I had a lab manager and a group of students, the annual tab to Brandeis grew to $100,000. I had an unpaid adjunct position with Brandeis and no research grant. Therefore The Alex Foundation had to pay the bills. Fund-raising became a constant refrain, and strain, in my life. But at least I had somewhere to continue my work.

Alex and Griffin moved to Brandeis in mid-January 2002, and Wart joined them shortly afterward. Wart had had a ball in New York City. My friend Maggie worked at home a lot, so he'd had her company as well as that of two young female Greys. He had been top bird for those five months; now he was going to have to get used to being at the bottom of the hierarchy again.

Space was again tight in the new quarters. The room measured no more than ten by fifteen feet. With three large cages, cupboards, bookshelves, a small refrigerator, a sink, T-stands for the birds, the lab manager's work desk with a computer, and a chair, it was crowded. Add one or two students and the stools they sat on during training, and, well, you get the picture. Fortunately, however, Arlene Levin-Rowe was my lab manager from late fall 2002 onward. Not only is Arlene a superb organizer and a natural with the birds, she is also one of the kindest and most even-tempered people I know. Life at the Brandeis lab would be hard to imagine without Arlene to make everything run harmoniously.

The cramped quarters affected the birds, particularly Alex. In Tucson, each bird had his own room, where he was trained and tested and where he slept. They shared the large communal lab space for chill-out times. Even at the Media Lab, where the communal room was small and crowded, the birds had had their own separate sleeping facilities. Now they shared one room for training, testing, chilling out, and sleeping. Alex, always the boss of the lab, became yet bossier. He was the unrivaled "big man on campus," and he made it known.

Alex would subject new students to an endless stream of requests—"Want corn," "Want nut," "Wanna go shoulder," and so on—effectively ensuring that the new person knew his lexicon. He'd always done that, but now there was more urgency. He also tried to trick new students into giving him extra corn in the afternoons, when he'd already had his noontime ration. His higher-octane bossiness was most obvious when we were trying to test Griffin on labels and concepts. In Tucson, Alex's opportunities to butt in were relatively rare; now they were constant. When Griff hesitated with his answer, Alex marched to the edge of his cage top and piped up with it from the back corner of the room. Alex occasionally even chimed in from inside his cardboard box on top of his cage.

If Griffin answered at all indistinctly, Alex would admonish him, "Say better." If I asked Griffin, "What color?" Alex might butt in with "No, you tell me what shape." Sometimes Alex gave the wrong answer, thus further confusing the already unsure Griffin. Alex was, to put it bluntly, a pain. Wart, meanwhile, was content to stay in his cage, playing enthusiastically with his toys.

The avian hierarchy that had always existed became even more sharply defined, with Alex insisting his seniority be properly acknowledged. He always had to be top bird. Literally. There are several photographs of the three birds and me in the new lab, always apparently portraying a harmonious "family." In reality, because Griffin liked to be on my shoulder, I had to arrange a more prominent perch for Alex, well in front of the other birds, but closest to my face—otherwise he refused to cooperate. Wart was usually on my hand, lowest of all. He was fine with that.

The first year at Brandeis was very difficult, and we didn't achieve a lot, mainly because I was distracted with administrative details and with job applications. Gradually, though, we settled into a productive routine. Because Alex was always butting in with Griffin, we decided to enlist him as one of Griffin's trainers, as we had attempted at Tucson. This he did enthusiastically. For the first time he was willing to question Griffin, something he had refused to do in Tucson.

Alex certainly tried to be helpful. At one point we were teaching Griffin the label "seven." Griffin gets very self-conscious when he can't produce what we want. His pupils get small. His body language broadcasts his discomfort. Sometimes he stops trying. Alex saw Griffin's difficulty and kept saying "sss," "sss," trying to prompt him. It was endearing, really. We hoped Griffin might learn faster with another Grey as a trainer. In the wild, after all, Greys learn vocalizations from each other. In fact, Griffin did make his first attempts faster after working with Alex, but then he had a more difficult time polishing his pronunciation.

It was sometimes amusing to figuratively step back and observe the two parrots: Alex saying, "What color?" and Griffin responding, "Blue," or whatever was appropriate. The fact that Griffin eventually spoke exactly like Alex— same tone, same inflection, same everything—added to the amusement.

We eased our way back into a work program by reviewing previously mastered tasks—colors, shapes, bigger/smaller comparisons, and so on. But then I embarked on what would turn out to be a remarkable series of studies with Alex in the challenging realm of numbers and mathematical concepts, some of which we had started in Tucson but then dropped. He was about to give new meaning to Woody Allen's line in *Annie Hall*: "Must be smart, went to Brandeis."

•   •   •

When we started the new series of number trials in the fall of 2003, Alex already knew the numbers one through six. He hadn't learned them in that order. He learned three and four first, as in "three-corner wood," for a triangular piece of wood, and "four-corner paper," for a square piece of paper. He later learned two, then five, then six, and lastly one. We now wanted to find out if Alex truly comprehended the number labels he used. When you ask a child below the age of three, "How many?" when holding up four objects, she will likely answer correctly, "Four." Ask the same child to give you four marbles from a dish of them, and she will just scoop up a handful and give them to you. As with words, number production does not necessarily mean comprehension.

Alex's test was quite straightforward. I would show him a tray on which there would be, for instance, two green keys, four blue keys, and six red keys, then ask, "What color four?" The correct answer, in this case, would be "Blue." Over a period of several days, Alex answered correctly in eight trials. I was impressed. Smart bird!

Suddenly he refused to continue with the trials for two weeks. He stared at the ceiling, gave responses of colors or object labels not on the tray, or fixated on an inappropriate label and repeated it endlessly. He preened. He gave every answer except the correct one. He asked for water or various foods, or said, "Wanna go back."

Then, for no obvious reason, he ended his strike in a remarkable way. I showed Alex a tray of objects, on which there were now two, three, and six blocks of different colors. I asked, "What color three?"

He answered very purposefully, "Five." Something was different about his attention and the tone of his response this time, unlike his previous indifference and inattention.

I asked again, "What color three?"

"Five," he shot back.

"No, Alex, what color three?" By now I was both puzzled and increasingly impatient. *Why say "five"? There is no set of five on the tray.*

"Five," he said again firmly.

Let's turn this around, I thought. I said to Alex, "OK, smarty pants. What color five?"

Without hesitation he said, "None."

I was astonished. *Is that what he meant?* Years earlier Alex had transferred the term "none" from a "same/different" study—telling us about the absence of a similarity or difference in shape, color, or material for a pair of objects—to a study on relative size. When first shown two objects of different colors but the same size and asked, "What color bigger?" he used "none" to signify that they were the same. He did it on his own, without training. Now, he apparently was using "none" to indicate the absence of a set of five objects, that is, using "none" to mean "zero," the absence of existence. To make sure this wasn't a fluke, we did six more trials in which trays lacked a set of one, two, three, four, and so on, and asked "What color one?" "What color two?" and so on. Of six trials, he was correct on five. His one error was to label a color not on the tray. Alex really did seem to have a concept of something like zero.

Who knows what went through his mind with that first series of "five" responses? He likely had become bored with the tests, which explains why he went on strike. After two weeks it was as if he said to himself, *OK, how can I make this interesting? I know—I'll label what's not on the tray.* Boredom is a powerful emotion for children in school and many adults. It surely isn't unique to humans.

Alex's use of "none" in this context is important for several reasons. First, zero is a highly abstract concept. A label for zero entered Western culture only in the 1600s. Second, Alex's use of "none" in this case was entirely his invention. We didn't teach him to do it. He figured it out for himself.

Shortly before I left Northwestern for Tucson, I had had a conversation with Tufts University philosopher Dan

Dennett. He wondered, "What if you asked Alex, 'What's green?' when there is nothing green on the tray? Would he say 'none'?" I was a little hesitant about pursuing this idea but tried it anyway. I showed a tray of different-colored objects to Alex and asked, "What's purple?" Nothing purple was on the tray. He looked at me and said, "Want grape." Grapes are purple.

He's outwitting me. He's not doing what I want him to do. He seems to be doing something clever. But how do I know when he's being clever and when he's wrong? This is harder than I supposed. I dropped the idea.

In the end, it was Alex who thought it up himself. This parrot, with his teeny brain, seems to have come up with a concept that had eluded the great Greek mathematician Euclid of Alexandria. Alex's use of "none" was as impressive as his leap to putting together the separate phonemes "nnn," "uh," and "tuh" to make a complete word. Probably more so. What would he do next?

In June 2004, the month we completed the number comprehension work, we started a study on addition. I hadn't planned such a study. Instead, it grew out of Alex's habit of butting into Griffin's sessions. We were teaching Griffin the number two by having him listen to two computer-generated clicks and asking, "How many?" Griffin didn't respond; he hunkered down and looked awkward. I generated two more clicks. "How many, Griffin?" No answer.

Then, from the top of his cage, Alex offered, "Four."

"Pipe down, Alex," I said testily. "I'm asking Griffin." I thought his response was random; after all, I had clicked twice.

Two more clicks. Still nothing from an increasingly anxious Griffin.

"Six," said Alex.

Six? Did he add up all the clicks and get six?

Psychologist Sally Boysen had investigated counting abilities and addition in chimps some time earlier, but she had used physical objects rather than sounds. I decided to do the same thing with Alex. Coincidentally, I had just received a one-year fellowship from the Radcliffe Institute, at Harvard, starting that fall. The fellowship was designed to allow me to collaborate with colleagues at Harvard who were studying number concepts in children.

We tested Alex's math skills by presenting him with a tray on which were two inverted plastic cups. Under one would be, say, two nuts, and under the other three nuts. We lifted the first cup and said, "Look, Alex," and then replaced the cup. We did the same with the second cup.

We then said, "How many total nuts?" Alex's accuracy in a series of such tests over the next six months was well above 85 percent. He indeed could add. This performance put Alex on a par with small children and chimpanzees.

What if there were no nuts under either cup and we asked, "How many total?" Would Alex say "none"? We tried it eight times. On the first four he said nothing. Instead, he looked at me as if he were thinking, *Hey, haven't you forgotten something?* He didn't say "two," as he might if he had confused the number of cups with what we actually wanted. On the next three trials he said "one." On the final test he again said nothing. Interestingly, in similar tests, chimps made the same "mistake" of saying "one."

Alex's performance indicated to me that his concept of zero was not as sophisticated as in humans. He didn't have "none" as the beginning of a number string, going up to six. When he responded "one," it was as if, like the apes, he knew he had to get as low in the number string as he could. (Later we started to train him on "none corner wood," for a circle, which he got.) But good as he was, he wasn't *that* good—somewhere between Euclid and the seventeenth century, so to speak.

Still, he was very good when it came to something called "equivalence." Again, he figured this out himself. No training.

Alex knew the vocal labels for Arabic numerals up to six. (We had started this work in the late 1990s, in Tucson, but returned to it in November 2004.) He also could label quantity for collections of objects, whether they were toy trucks, keys, or wooden cubes, and so on, again up to six. But we had never paired the Arabic numerals with sets of objects. The question was, did he understand that the squiggle that is the Arabic numeral six represents sixness? This is what is meant by equivalence. We also weren't sure that Alex knew that the numeral six is bigger than five, five is bigger than four, and so on. Alex had not learned his numbers in order, as children do. Learning numbers in the proper order implies increasing quantity. Could he overcome this disadvantage?

Now, in a trial of one type of test, we would place a green plastic Arabic numeral five next to three blue wooden blocks and ask, "What color bigger?" Physically, the collection of wooden blocks was bigger than the Arabic numeral. If Alex were guided only by physical size, he would have said "blue." He didn't. He said "green." To a very high degree of accuracy he repeatedly judged the question according to number. In another series of test trials, we showed him two Arabic numerals of different value and different color. Again we asked, "What color bigger?" Again he almost always got the right answer. We hadn't trained Alex to do this. He figured out for himself that the Arabic numeral six represents six somethings, the numeral five, five somethings, and so on. And he knew that six is bigger than five, and so on down the number string. Chimpanzees cannot do this without extensive training.

These are truly sophisticated numerical abilities, once assumed to be the sole province of the human brain, and supposedly made possible by human language. Once again, Alex had done what he was not supposed to do.

Mike Tomasello is a brilliant primatologist at the Max Planck Institute for Evolutionary Anthropology, in Leipzig, Germany, and a good friend. His specialty is the evolutionary origin of certain higher cognitive functions in humans, including language. We often laugh about how he sometimes finishes his talks at scientific meetings. Like most of his colleagues, Mike believes that all the scientific evidence points to the fact that these "higher" functions in humans arose uniquely from the primate brain, and states as much at the end of his talks. But then he often throws up his hands and says with humor and frustration, "Except for that damn bird!" Alex.

The media loved the "none" and "equivalence" stories, particularly the zerolike concept. Comparisons with Europeans' late invention of a label for zero made it an easy mark. I felt that Alex's recognition of equivalence deserved more attention than it got, however, because it demonstrated a degree of abstraction and cognitive processing that even I had not imagined possible. It was increasingly clear to me that our future together was going to produce amazing achievements that would make Alex's previous few decades look intellectually pedestrian by comparison.

My glorious year of intellectual freedom and financial security on the Radcliffe Fellowship came to an end in the summer of 2005. I was uncovering cognitive abilities in Alex that no one believed were possible, and challenging science's deepest assumptions about the origin of human cognitive abilities. And yet I was without a job. I was also without a grant. I had to apply for unemployment insurance. I ate fourteen tofu meals a week, and I kept my thermostat at 57 degrees during the winter to minimize household expenses. It was only thanks to generous donors to The Alex Foundation that I could keep working with Alex.

The media portrayed Alex as a brainiac, and he was—a bird brainiac. But there were many more sides to Mr. A. than just his cognitive achievements. He was bossy and obstinate. He was playful, not just with toys but intellectually, when he deliberately gave wrong answers. He was mischievous and affectionate. And he projected a sense that, while dependent on us for his material needs, he was supremely confident in who he was as an individual. He owned us as much as we owned him. Overall, he was a Puck-like character.

A favorite excursion was to a small lobby close to the lab. "Go see tree," he would say at least two or three times a week. Students would follow his orders and take him to the lobby. They usually took his perch, too. But Alex much preferred to sit on the back of a small couch next to the window. He liked to look at birds in the tree close to the window, and at trucks that went by on the road below. Students passing on the staircase below the window were oblivious of Alex's rapt attention to their comings and goings and to the cheerful whistles he produced for their benefit. He liked to wolf-whistle at boys who walked through the lobby, much to the consternation of the girl students tending him.

His favorite lobby activity, though, was to dance to "California Dreamin'," which the students rendered enthusiastically on every visit. That tradition started a few years back when someone was playing the Mamas and Papas version in the lab, and Alex began bobbing his head vigorously along with the music. After that he head-bobbed to the students' version, especially after Arlene taught everyone the lyrics.

We had lab schedules, of course: meal times and work times. But Alex had his own daily activities. After lunch of warm grain, he often liked to retreat to the top of his cage or inside his cardboard box. Eyes half closed, he entered into a monologue of words and phrases: "Good boy . . . Go eat dinner . . . You be good . . . What's your problem? . . . C'mon." He did the same thing at around four-thirty each day: "Wanna go chair . . . What color? . . . Shou-wah." Arlene called these monologues the "Alex Chronicles," his ruminations on events of the day. Sometimes he'd practice a new label, so, for example, we could track how he approached "seven" as "s . . . one," then "s . . . none," then "seben."

The birds always had the company of a couple of students and Arlene during most of the day. I usually arrived in the late afternoon. And there were occasional visitors, sometimes quite distinguished. One such was the Canadian novelist Margaret Atwood. A few years back a copy of *Oryx and Crake* had arrived on my desk, with no explanation. The book is Atwood's fantasy about humanity's final days. I realized why I'd been sent a copy when I came across a passage in which a boy, Jimmy, is watching old television footage of a Grey parrot that can identify colors, shapes, and numbers, and uses the term "cork-nut" for almonds. That was Alex, of course. Back in the Purdue days, Alex had said "cork" when I first showed him an almond. That was reasonable, because superficially an almond shell resembles cork. We therefore started to use "cork nut" for almonds, which Alex learned.

Shortly after I received Atwood's novel, I heard that she was to be at The Radcliffe Institute to receive its annual gold medal. I thought she might enjoy meeting the real Alex, so I contacted her publicist and suggested a visit. I picked her up from Radcliffe and drove her to Brandeis. She was elegantly dressed, friendly but reserved. For whatever reason, however, Alex decided to be completely uncooperative. I tried for fully twenty minutes to get him to say "cork nut." Nothing. When he finally did condescend to utter a word, it was "walnut . . . walnut."

Frustrated and apologetic, I turned to Griffin, who loves almonds and was, I thought, certain to say "cork nut," as he so often does with great alacrity. Yet all he would say was "walnut . . . walnut." Eventually Atwood's driver arrived, and she left, politely thanking me for the visit. She was barely through the door before both Alex and Griffin were impishly piping up, "Cork nut . . . cork nut . . . cork nut."

I have my own cork nut story. I once went into Trader Joe's and asked a sales assistant where I could find the cork nuts. The guy looked at me as if I were crazy. It took me a few seconds to realize what his problem was. "Oh, almonds, I mean almonds," I said, embarrassed. "Cork nuts is what my child calls them." I turned and scooted out of sight as fast as I could. It is very easy to lapse into the language of the lab. The students do it all the time, mimicking Alex's particular cadence: "Shou-wah . . . foo-wah . . . thu-reeh." It's an in-joke that occasionally escapes to the outside world.

I've said earlier that Alex preferred guys to girls and expressed his special likes by performing the Grey's mating dance. During the early part of 2007, Alex became hypersexual with his favorites. Poor Steve Patriarco. For about six months, whenever Steve picked up Alex, he raced up to Steve's shoulder, where he puffed up his feathers, danced from foot to foot, and regurgitated food. It got to be quite ridiculous. Alex was also distinctly uninterested in working during this period.

On the advice of our vet, we took away Alex's cardboard box. Ever since Tucson, Alex had enthusiastically chewed windows and doors in cardboard boxes. He loved to spend time in his "house," chilling out, producing monologues, and commenting on activities in the lab. The box was a nest substitute. The vet thought that it might be exacerbating whatever hormones were raging within Alex. We also fed Alex tofu to temper them.

By August he was stable again, working again, and no longer quite as frequently performing his mating dance for Steve. We gave him back his box. One of the students brought birthday cake into the lab at about that time, and we all shared it, as usual, including the birds. "Yummy bread," Alex said appreciatively. He had known "yummy" previously, and "bread." But "yummy bread" was his creation.

At the end of August, workmen cut down the tree just outside the window in the lobby. Alex could no longer watch the birds.

I had been thinking about doing work on optical illusions with Alex ever since I was at the Media Lab. In the summer of 2005, I teamed up with Patrick Cavanagh, a psychology professor at Harvard, to put the idea into practice. The human brain plays many tricks on us, so we sometimes see things not as they are. Patrick and I planned to ask a simple but profound question: does Alex *literally* see the world as we do? That is, does his brain experience the optical illusions just as our brains do?

I envisaged this work as the next horizon in my journey with Alex, beyond naming objects or categories or numbers. Bird and human brains diverged evolutionarily some 280 million years ago. Does that mean that bird and mammalian brains are so different structurally that they operate very differently, too?

Until a landmark paper by Eric Jarvis and colleagues in 2005, the answer to this question had been a resounding *yes!* Look at a mammalian brain and you are struck by the multiple folds of the massive cerebral cortex. Bird brains, it was said, don't have such a cortex. Hence, their cognitive capacity should be extremely limited. This, essentially, was the argument I had faced through three decades of work with Alex. He was not supposed to be able to name objects and categories, understand "bigger" and "smaller," "same" and "different," because his was a bird brain. But, of course, Alex did do such things. I knew that Alex was proving a profound truth: brains may look different, and there may be a spectrum of ability that is determined by anatomical details, but brains and intelligence are a universally shared trait in nature—the capacity varies, but the building blocks are the same.

By the turn of the millennium, my argument was beginning to gain ground. It wasn't just my work with Alex but others' work, too. Animals were being granted a greater degree of intelligence than had been previously allowed. One sign of this was that I was asked to co-chair a symposium at the 2002 annual meeting of the American

Association for the Advancement of Science, called "Avian Cognition: When Being Called 'Bird Brain' Is a Compliment." The preamble read as follows: "This symposium demonstrates that many avian species, despite brain architectures that lack much cortical structure and evolutionary histories and that differ so greatly from those of humans, equal and sometimes surpass humans with respect to various cognitive tasks." Even five years earlier, such a symposium would have been a difficult sell. That was progress. Jarvis's paper three years later effectively said that bird and mammalian brains are not so very structurally different after all. More progress.

When Patrick and I submitted our grant proposal to the National Science Foundation in July 2006, we were expecting that, in some respects at least, Alex would see our world as we do. We didn't wait to hear whether we would be funded before we embarked on some preliminary work. We chose a well-known illusion as the first test. You have probably seen it in psychology textbooks and popular articles: two parallel lines of equal length, both with arrows at the ends, one with the arrows pointing out, the other with the arrows pointing in. Despite being the same length, the line with the arrows pointing in looks longer to human eyes. That's the illusion. We had to modify the test a little so as to use Alex's unique abilities; we varied the color of the two lines, keeping the arrows black. We then asked, "What color bigger/smaller?" Right away, and repeatedly, Alex selected the one that you or I would choose. He did see the world as we do, at least with this illusion. That was a very promising step.

By June 2007, Patrick and I were pretty sure that we would get our grant, and by the end of August we learned that it would start on September 1, a Saturday. We would have money for a year. The following Monday we threw a party to celebrate, on the seventh floor of Harvard's William James Hall. I was especially happy, and relieved to see my financial woes lessen.

I'd been teaching part-time at Harvard's Extension School since 2006 and in the psychology department beginning in 2007. I survived with a little extra income from The Alex Foundation, but it still had meant tofu and a 57-degree thermostat setting for me. The new grant would change all that. I would become a regular research associate, with a small but decent salary and benefits. And 35 percent of the lab costs would be covered. That was $35,000 less I would have to raise for that year. I could not have been happier. It wasn't a tenured professorship, but it was quite an improvement.

Alex was a little subdued that week, though nothing out of the ordinary. The birds had had some kind of infection the previous month, but they were now fine. The vet had given them all a clean bill of health. On the afternoon of Wednesday the fifth, Adena Schachner joined me and Alex in the lab. She is a graduate student in the psychology department at Harvard, researching the origins of musical abilities. We thought it would be interesting to do some work with Alex. That evening, we wanted to see what types of music engaged him. Adena played some eighties disco, and Alex had a good time, bobbing his head in time with the beat. Adena and I danced to some of the songs while Alex bobbed along with us. Next time, we promised ourselves, we would get more serious about the music work.

The following day, Thursday the sixth, Alex wasn't much interested in working on phonemes with two of the students during the morning session. "Alex very uncooperative in the task. Turned around," they wrote in Alex's work log. By midafternoon he was much more engaged, this time with a simple task of correctly selecting a colored cup, underneath which was a nut. I arrived at five o'clock, as usual. Arlene had left for the day. She and the students had already moved the floor mats to one side for the regular Friday morning cleaning by the maintenance crew. Shannon Cabell, a student, was with me. We sat at the computer, with Alex between us on his perch, looking at the screen. I was working on new optical illusion tests, trying to get the colors and shapes right—nothing demanding, just fiddling with things. Alex was affectionate and chatty as usual.

At six forty-five the supplemental lights went on, as usual, a signal that we had a few minutes left to clean up. Then the main lights went off, and it was time to put the birds in their cages: Wart first, then Alex, then the always reluctant Griffin.

"You be good. I love you," Alex said to me.

"I love you, too," I replied.

"You'll be in tomorrow?"

"Yes," I said, "I'll be in tomorrow." That was our usual parting exchange. Griffin and Wart said nothing, as usual.

I left and drove forty minutes to my home in Swampscott, on the North Shore. I went through e-mail, had a bite to eat and a glass of wine, and went to bed.

The following morning I was up by six-thirty, as usual. After showering and stretching, I took a walk by the ocean, which I loved doing each day. It was an important reason why I chose to live where I do. The sun was already way up but still low enough to project scintillating tracks across the calm ocean. It was one of those glorious crystal-blue early September New England days. Captivating.

I was back at the house eating breakfast in front of my computer by eight-thirty. There was an e-mail message waiting for me. "This is to confirm that we have been successful with the ITALK grant application," the message read. "You are one of our consultants. Congratulations! We'll be contacting you again soon." The message was from a colleague in Europe. He was part of a consortium of researchers that had proposed a major project on the evolution of language involving computer models and robotics. The project had been ranked first out of thirty-two competing proposals and was awarded six million euros, to begin in February 2008. Although I wasn't going to be an active member of the research team, I would be flying to Europe at least once a year to brainstorm about results and ideas.

Coming just days after the formal approval of our NSF grant, this European news was thrilling, a bonus really. I jabbed the air with both fists and declared out loud, "Yes! Things are really turning round at last!" I immediately wrote back to my colleague, got up, went to the kitchen, and poured myself another cup of coffee.

As I stood quietly there for a few minutes, savoring the coffee's rich aroma, a thought crossed my mind, as it did from time to time, something that my friend Jeannie once said: had I gotten a different Grey that day back in 1977, Alex might have spent his life, unknown and unheralded, in someone's spare bedroom. I didn't, of course, and here we were with a history of astonishing achievements behind us, and poised to journey to the next horizon and beyond in our work together. And we had the resources we needed. I allowed myself to savor all this, too, a sense of happiness, excitement, and security that had eluded me since the heady days at the Media Lab. *Yes!* I then returned to my computer.

In the interim, another e-mail had arrived. In the subject line was a single word: "Sadness." My blood turned to ice as I read the message. "I'm saddened to report that one of the parrots was found dead in the bottom of his cage this morning when Jose went to clean the room . . . not sure which? . . . in the back left corner of the room." It was from K.C. Hayes, the chief veterinarian in the animal care facility at Brandeis.

I was in raw panic. *No . . . no . . . no! Back left corner of the room. That's Alex's cage!* I was gasping for breath, trying to stave off rising terror. *Maybe he's mixing up his right and his left. Maybe he made a mistake. Maybe it's not Alex. It can't be Alex!* Even though I clung to that feeble hope as I snatched at the phone, I knew K.C. had not made a mistake. I knew Alex was dead. Even before I could dial, a second e-mail from K.C. chimed onto the screen. The message was simple. "I'm afraid it is Alex."

I reached K.C., barely able to speak through my tears and pain. He told me he had wrapped Alex in a piece of cloth and put him in the walk-in cooler, down the hall from the lab. I threw on jeans and a shirt and jumped in my car. I'll never know how I managed to drive, given my state. I called Arlene, because I didn't want her to walk into the lab unprepared. She was just driving into the parking lot down the hill from the lab when I reached her. "Alex is dead, Alex is dead," I wailed. "But maybe, maybe they made a mistake. Maybe it isn't Alex. Please go find out, Arlene." What was I saying? I knew K.C. hadn't made a mistake. I knew Alex was dead. But I said those words anyway, as if they might make untrue what I knew to be true.

Poor Arlene. She was crying hysterically now, too. Eventually she said she'd go to the lab and find out what had happened. She ran up the slope and in through the side door to the facility. She got to the lab just moments after our friend and lab volunteer Betsy Lindsay had arrived. Betsy hadn't even noticed anything amiss. Arlene saw immediately what she, too, had hoped against all reason she would not see: Griffin and Wart were in their cages, doors closed. The door to Alex's cage was open slightly. The cage was empty.

When I arrived in the lab almost an hour later, Arlene and I held each other and sobbed for quite some time. Wave after wave of pain and despair washed over us, a torrent of shared disbelief. "Alex can't be dead," Arlene whispered through her tears. "He was larger than life."

We knew that we had to take Alex to the vet for an autopsy, but neither of us could bring ourselves to go fetch him from the cooler. Betsy did that for us and put him in a small carrying case. We decided that Arlene was the more capable of the two of us to make the forty-minute drive. She had done it many times, when Alex or one of the other birds needed treatment or a checkup. This time would be different. This time we wouldn't be bringing Alex back with us.

Karen Holmes, one of the veterinarians at the practice, greeted us with hugs of sympathy. She led us to the mourning room, where we placed Alex, still wrapped, still in his carrier, next to us on the sofa. Arlene and I sat holding hands, crying, not saying much. Karen asked if I wanted to see Alex one last time, but I didn't. Years ago I had seen my father-in-law in his casket. For a very long time I couldn't dispel the image of him lying there, drained of life. I determined then never to look at death again, and I stuck to that determination, even when my mother died.

I wanted to remember the Alex I'd put in the cage the previous night. Alex, full of life and mischief. Alex, who had been my friend and colleague for so many years. Alex, who had amazed the world of science, doing so many things he was not supposed to be able to do. Now he had died when he was not supposed to, two decades before the end of his expected life span. *Damn you, Alex.*

I wanted to remember the Alex whose last words to me were, "You be good. I love you."

I stood, put my hand on the door, and whispered, "Goodbye, little friend." I turned and walked out of the clinic.

Chapter 9

||||||||||||||||||||||||

What Alex
Taught Me

Alex left us as a magician might exit the stage: a blinding flash, a cloud of smoke, and the weaver of wizardry is gone, leaving us awestruck at what we'd seen, and wondering what other secrets remained hidden. Alex's sudden, unexpected departure left me in awe at his achievements and wondering what else he would have done had he stayed. He left at the height of his powers. To some what he did seemed magical, or at least otherworldly. Indeed, he had given us a glimpse of another world, one that had always existed but remained beyond our view: the world of animal minds. I barely had a voice when I was a child. Yet this powerful little feathered presence gave voice to a hidden world of nature. He was a great teacher to me and to us all.

The greatest single practical lesson Alex taught me was patience. Ever since childhood I have been a determined kind of person. Whatever I wanted to do, I would pursue doggedly and see through to the finish. I embarked on The Alex Project in the early seventies with the same grit and idealism I always have, or rather had. I wonder if I might have had second thoughts about the venture had I known the multitude of practical obstacles and the anti-bird-brain prejudices I would face over the years. But I didn't know, of course, and in any case I doubt that anything would have been capable of deflecting me, so convinced was I of an amazing world of animal cognition to be explored. But oh, what patience it took to get us to where we were when he died.

Scientifically speaking, the single greatest lesson Alex taught me, taught all of us, is that animal minds are a great deal more like human minds than the vast majority of behavioral scientists believed—or, more importantly, were even prepared to concede might be remotely possible. Now, I am not saying that animals are miniature humans with somewhat lower-octane mental powers, although when Alex strutted around the lab and gave orders to all and sundry, he gave the appearance of being a feathery Napoleon. Yet animals are far more than the mindless automatons that mainstream science held them to be for so long. Alex taught us how little we know about animal minds and how much more there is to discover. This insight has profound implications, philosophically, sociologically, and practically. It affects our view of the species *Homo sapiens* and its place in nature.

Exactly how scientists came to espouse ideas about animal minds that were so at odds with what nonscientists would call common sense is fascinating and instructive. It bears exploring, because it tells us a lot about ourselves as a species. Humans have always tried to make sense of their world and their place in it. Foraging people, living in close harmony with nature and her rhythms, see themselves as closely connected to other living things in their worlds. They see themselves as an integral part of the whole of nature. We see this expressed in the mythologies and folk tales of Australian Aborigines and Native Americans, for instance. The same would have been true for all populations of *Homo sapiens* through the six thousand generations after which our species arose, and until relatively recently in human history. When Western civilization began to take root with the Greeks, a very different way of thinking began to emerge.

Aristotle, in the fourth century B.C.E., constructed a

view of the natural world that is, in its essence, still with us. He ordered all living and nonliving things on a ladder of perceived importance based on mind. Humans were at the top, below the gods, a place earned by our great intellect. On lower and lower rungs were the lesser creatures, and finally the plants; lowest of all was the mineral world. The Judeo-Christian tradition enthusiastically adopted Aristotle's blueprint, in which humans were given dominion over all living things and the earth. This description of nature came to be known as the Great Chain of Being. Humans were not only different from all other of God's creatures, but also distinctly superior.

Little changed when Darwin argued that we are the product of evolution rather than of God's creation. The Great Chain of Being, a static ordering of life, merely morphed into a dynamic process of progressive evolution. Simple forms yielded more complex forms over evolutionary time, ultimately giving rise to humans as the pinnacle and goal. (Darwin did not put it this way, but anthropocentric others had no trouble interpreting his theory in this fashion.) All other living things were for our exploitation. We still were different from and superior to the rest of nature, despite our connection to nature through our evolutionary heritage. Or so most scientists believed until not so very long ago. Vanity, thy name is *Homo sapiens*.

What Alex Taught Me

Recognizing that *Homo sapiens* is connected to the rest of nature through evolution had hurt the human psyche. The belief that our intellect, and particularly spoken language, was unique in nature was a life vest for our damaged pride. It kept us afloat above lower creatures. As Thomas Henry Huxley, Darwin's champion, wrote in his 1872 book, *Man's Place in Nature*: "No one is more strongly convinced than I am of the vastness of the gulf between . . . man and the brutes, for he alone possesses the marvelous endowment of intelligible and rational speech [and] . . . stands raised upon it as on a mountain top, far above the level of his humble fellows."

This lofty sentiment changed little with the passage of a century. Norman Malcolm's 1973 presidential address to the American Philosophical Association said essentially the same thing: "The relationship between language and thought must be . . . so close that it is really senseless to conjecture that people may *not* have thoughts, and also really senseless to conjecture that animals *may* have thoughts." Malcolm said this just a year before I decided to embark upon what would become The Alex Project—and *after* the Gardners had published their first paper on Washoe. But the equation was simple and, for many, conclusive: language is necessary for thought; animals don't have language; therefore, animals don't think. This also was the behaviorists' gospel, a movement that began in the 1920s and was still dominant when I started working with Alex. Animals are automatons, responding mindlessly to stimuli, the behaviorists' contended, exactly the same position posited by René Descartes three and a half centuries earlier.

Little wonder, then, that passions ran so high at the Clever Hans conference I discussed earlier: people working with apes and dolphins were challenging cherished notions about supposed human uniqueness. Genuine methodological issues needed to be aired about the ape-language work. But the underlying drive of the conference had been a desire to protect the assumption of human primacy. That assumption had never really been tested.

Nevertheless, the fortress of human uniqueness came under attack in the 1980s and began to crumble. We once thought only humans used tools; not so, as Jane Goodall discovered her chimps using sticks and leaves as tools. OK, only humans *make* tools; again not so, as Goodall and later others discovered. Only humans had language; yes, but elements of language had been discovered in nonhuman mammals. Each time nonhuman animals were found doing what was the supposed province of humans, defenders of the "humans are unique" doctrine moved the goalposts.

Eventually, these defenders conceded that evolutionary roots of certain cherished human cognitive abilities could indeed be found in nonhuman animals, but only in large-brained mammals, particularly in apes. By doing the things he did, Alex taught us that this, too, was untrue. A nonprimate, nonmammal creature with a walnut-sized brain could learn elements of communication at least as well as chimps. This new channel of communication opened a window onto Alex's mind, revealing to me and to all of us the sophisticated information processing— thinking—I described in earlier chapters.

By implication, a vast world of animal cognition exists out there, not just in African Grey parrots but in other creatures, too. It is a world largely untapped by science. Clearly, animals know more than we think, and think a great deal more than we know. That, essentially, was what Alex (and a growing number of research projects) taught us. He taught us that our vanity had blinded us to the true nature of minds, animal and human; that so much more is to be learned about animal minds than received doctrine allowed. No wonder Alex and I faced so much flak!

We faced a flurry of goalpost moving, too. Birds can't learn to label objects, they said. Alex did. OK, birds can't learn to generalize. Alex did. All right, but they can't learn concepts. Alex did. Well, they certainly can't understand "same" versus "different." Alex did. And on and on. Alex was teaching these skeptics about the extent of animal minds, but they were slow, reluctant learners.

Science has to be rigorous in its methodology. I understand that. It's why I worked so painstakingly over the years. It's why I insisted that we test Alex through so many repetitive trials before we could say with statistical confidence that he did indeed have this or that cognitive ability. Poor bird. No wonder he'd sometimes get bored and refuse to cooperate, or play creative tricks on me. No wonder from time to time Alex pushed me to go beyond the task at hand. When he "spelled" "Nnn . . . uh . . . tuh" in exasperation that I had not given him a nut, he went beyond anything I had asked. When he got me to ask a question so he could answer "none," he was applying that concept to a novel context.

What did these and other things that Alex had done teach me? To believe that Alex had a degree of consciousness that even less radical behaviorists would flatly deny. Can I prove it, the way I proved Alex could label objects and learn concepts? No, I can't. Although language is no longer widely held to be a prerequisite for thinking—I often think visually, for instance, as many people do, and nonhuman animals might do this, too—language is necessary to *prove* another individual is conscious. Language allows us to explore the workings of another individual's mind as nothing else can. If I could have asked Alex, "Why did you chew up that grant application back at Purdue?" or "What were you thinking when you chewed up the slides in my desk at Northwestern?" and had him reply, "Oh, I was just having fun," or "I knew it would get your goat," then I'd have glimpsed his consciousness. But Alex didn't have the use of language the way you and I do. So I can't prove he had a degree of consciousness. But the way he behaved surely was suggestive.

Alex taught me to believe that his little bird brain was conscious in some manner, that is, capable of intention. By extrapolation, Alex taught me that we live in a world populated by thinking, conscious creatures. Not human thinking. Not human consciousness. But not mindless automatons sleepwalking through their lives, either.

Some people take this new understanding of animal minds as an argument for treating animals as if they had the same rights we give to ourselves. That is as wrong as the behaviorists' restrictive gospel. Parrots and other pets are not little humans. They are their own beings. Do they deserve to be treated with care and kindness? Of course. As an intelligent flock animal, Greys need a lot of companionship, and it would be cruel to adopt one as a pet only to leave it alone all day. But that doesn't mean Greys or other animals have wide-ranging political rights.

The most profound lesson that Alex taught us concerns the place of *Homo sapiens* in nature. The revolution in animal cognition of which Alex was an important part teaches us that humans are not unique, as we long believed. We are not superior to all other beings in nature. The idea of humans' separateness from the rest of nature is no longer tenable. Alex taught us that we are *a part of* nature, not *apart from* nature. The "separateness" notion was a dangerous illusion that gave us permission to exploit every aspect of the natural world—animal, plant, mineral—without consequences. We are now facing those consequences: poverty, starvation, and climate change, for example.

My ecologist friends are more aware than most scientists of the interconnectedness of living things in the world, and their dependence on the nonliving realm. But even this awareness is relatively new in appreciating the complexity of animal and plant communities, locally, regionally, and globally. For much of the twentieth century, all sciences, including biology, were obsessed with reductionism: viewing the world at all levels, from the smallest to the largest, as merely a machine made of parts. Take the machine apart, examine the individual pieces, and we would understand how the world works.

Reductionism has had many triumphs in understanding the nature of the parts and how some parts fit together. It enabled us to build computers and devise powerful medicines, for example. But some scientists admit that reductionism falls short of its ultimate goal: understanding how the world works. It falls short because it fails to recognize the connectedness, the unity, that is the deep essence of nature in all realms. Not in the sense of physicists seeking the ultimate fundamental particle or the theory of everything. There is a oneness in nature in the sense of *interdependence*.

My scientifically literate, nonscientist friends get this idea instantly and intuitively. It "feels right," if you will. Deb Rivel, a friend and The Alex Foundation board member, put it this way: "Alex taught me the meaning of oneness. What I learned from him also supported what I always have known to be true: that there is just one Creation, one Nature, one good, full, complete Idea, made up of individuals of all shapes and designs, all expressing their oneness with one God. We are not different because we look different, but we all reflect the eternal beauty and intelligence of one Creation in our own peculiar way. It's what makes up the whole—this textured fabric of thought and existence—and knowing Alex has underscored to me how much the same we really are."

Deb expresses beautifully what people who believe in a God say they learned from Alex. I personally am not much interested in organized religion. But I strongly believe in the oneness and beauty of the world that Alex taught Deb and taught me. My philosophy of life is based in an appreciation of the holistic nature of the world, seeds of which were planted in my childhood, put there by No-Name and the love of nature to which that bird led me. My "religion" is therefore much more akin to the Native American lore of being and belonging, of equality with and responsibility for nature. Who knows what other amazing things we might have seen through our window onto Alex's mind had he stayed? In any case, he did leave me this great gift of what was once known and embraced but was lost: the oneness of nature and our part in it.

Alex didn't stay, and his passing taught me the true depth of our shared connection. The searing pain and grief I experienced when he died taught me that. I had always loved the little guy, of course, as one does an individual with whom one works so closely day after day over three decades. He had depended completely on me and my students for his material needs, but he always had an air of independence about him, haughty independence. And I had kept my true attachment to him in close check, so much so that it became invisible even to me. No longer.

I took care of Alex, as any dutiful Grey owner would, but he was such a free spirit that I never felt I *owned* him. This feeling is best expressed in one of my favorite films, *Out of Africa*. Based on Isak Dinesen's memoir of the same name, it tells the wrenching story of doomed love between Danish baroness Karen Blixen (Dinesen's real name) and Denys Finch-Hatton, a dashing hunter and aviator, set in the mystical Ngong Hills of southwest Kenya. The book opens with the simple and yet deeply evocative phrase, "I had a farm in Africa."

It's difficult to explain, but when you go to Africa, the place gets under your skin, burrows into your soul. And so that simplest of opening lines instantly taps into the most fundamental of emotions. It also stirs a deep sorrow that comes from knowing the devastation that is now being visited upon much of these primal lands, victim of the double depredation of limitless greed and desperate need. Sad. Where's the recognition of oneness here?

One attraction of the story for me involves a certain identification with this woman and her quest in life. She,

Acknowledgments

||||||||||||||||||||||||

To all those who sent e-mail, snail mail, and phoned after Alex died, who convinced me of the need for this book . . . to Arlene, without whom I couldn't have survived . . . to all those who provided financial assistance for the research over the years, whether a few dollars or several thousands, or by spending countless hours arranging fundraisers . . . to all those who gave their emotional support through thick and thin . . . thank you! I would also like to acknowledge Roger Lewin's significant help in drafting the manuscript.

Index

Index

Index

Index

Index